Past climatic and environmental change is of prime importance in understanding climatic changes of today. This book describes and discusses the great environmental changes revealed by a study of a small area in central Norfolk, which has given a remarkable wealth of data concerning the many consequences of climatic change over the past few hundred thousand years. Rather than a theoretical treatment of climatic changes, this book is a unique 'case study' of an investigation of past climatic change.

Pleistocene palaeoecology of central Norfolk

a study of environments through time

Pleistocene palaeoecology of central Norfolk

a study of environments through time

R. G. WEST

Professor of Botany, University of Cambridge
Fellow of Clare College, Cambridge

The right of the
University of Cambridge
to print and sell
all manner of books
was granted by
Henry VIII in 1534.
The University has printed
and published continuously
since 1584.

CAMBRIDGE UNIVERSITY PRESS
CAMBRIDGE
NEW YORK   PORT CHESTER   MELBOURNE   SYDNEY

CAMBRIDGE UNIVERSITY PRESS
Cambridge, New York, Melbourne, Madrid, Cape Town, Singapore, São Paulo, Delhi

Cambridge University Press
The Edinburgh Building, Cambridge CB2 8RU, UK

Published in the United States of America by Cambridge University Press, New York

www.cambridge.org
Information on this title: www.cambridge.org/9780521116091

First published 1991
This digitally printed version 2009

A catalogue record for this publication is available from the British Library

Library of Congress Cataloguing in Publication data

West, R. G.
Pleistocene palaeoecology of central Norfolk : a study of
environments through time / R. G. West
p.   cm.
ISBN 0 521 40368 5
1. Geology, stratigraphic–Pleistocene. 2. Geology–England–
Norfolk. 3. Palaeoecology–England–Norfolk. I. Title
QE697.W528   1991
560′.45′094261–dc20   90–21856   CIP

ISBN 978-0-521-40368-9 hardback
ISBN 978-0-521-11609-1 paperback

CONTENTS

PREFACE AND
ACKNOWLEDGEMENTS

The research described here started in 1964 and ended in 1989. What began as a study of a section showing organic sediments of Ipswichian (last) temperate Stage age, expanded as the sand and gravel workings at Roosting Hill were extended. Devensian stadial (full-glacial) and interstadial sediments were exposed, and later, in 1983, sediments of an earlier temperate stage were found, resting on glacigenic sediments of the major (Anglian) ice advance which covered East Anglia. To find such a plethora of evidence of Pleistocene environmental changes near the surface in a relatively small area is exceptional. I hope the problem of organising the reporting and discussion of the evidence and its complexity has been helped by the treatment I have followed: outlining the major stages represented, treating each stage separately, with additional discussion of more general points at the end.

I gratefully acknowledge the very generous help I have received from many sources in this study. The Natural Environment Research Council supported the work by grants for assistance with pollen analysis, analysis of macroscopic remains and excavations. I am particularly indebted to Miss R. Andrew for her major contribution to the analysis of pollen assemblages and likewise to Mrs M. Pettit for the analysis of macroscopic plant remains. Dr L. Phillips also contributed certain pollen analyses. The owners of the Roosting Hill sand and gravel pits, Barker Bros., freely gave permission for examination of sections, excavations and borehole records, and I thank Mr T. Barker, Mr M. Eves, Mr H. Mobbs, Mr M. Moore, Mr M. Bellboddy and Mr S. Sadler for their interest and ready cooperation in the field investigations. Mr R.A.D. Markham first informed me of the interest of the site. Mr J. Webb kindly gave me notes on certain sections he had seen. The several local farmers, in particular Mr Gow, allowed me access to their land. Numerous colleagues assisted with excavations, in particular the group of 18 who descended on Beetley in the winter of 1983. I am indebted to Dr P.L. Gibbard for numerous fruitful discussions on stratigraphical problems. Dr V.R. Switsur provided advice on radiocarbon dating and facilitated the dating. Mrs S.M. Peglar and Mrs J. Dye gave much help with preparations and sediment analyses. Mr C. Auton of the British Geological Survey provided local geological information. Mrs H. West and Miss P. Gristwood gave invaluable assistance in numerous levelling expeditions. The RAF aerial photographs are reproduced by permission of the Controller of Her Majesty's Stationery Office.

CHAPTER ONE

Introduction

The Pleistocene is the latest part of the geological record, stretching back some 2.4 million years. It is characterised by much evidence of environmental changes, for example climatic and sea-level changes. Such changes are expressed in the record of changing sediments during the period, and in the record of changing plant and animal assemblages found fossil in the sediments. As a result, we have a reasonably clear idea of the magnitude of changes in north-west Europe, with the fluctuating climates of the Pleistocene leading, in the later part of the period, to alternating temperate and cold stages, the latter including times of great development of ice sheets.

The detail of evidence available in the Pleistocene for the reconstruction of past climates, flora and fauna is remarkable, the more so because the physical processes and the species concerned are seen to be clearly connected wih the present-day world. Indeed, the origin of present-day conditions must be sought in the geologically recent times of the Pleistocene. For those interested in environmental problems the Pleistocene record of change is an essential background.

The Pleistocene sediments of East Anglia offer the most detailed record of environmental and biota history in the British Isles during this time. The reason for this is that it is a low-lying area, partly overlapping the marine North Sea basin to the east, and is in the marginal region of the great ice advances of the later part of the Pleistocene. The result is that there is a record in the marine sediments of the lower part of the Pleistocene, and a record of alternating temperate stages and cold stages associated with ice advances in the later part of the Pleistocene. This is not to say that the record is complete. If the expression of climatic change in the deep sea record (Shackleton & Opdyke, 1973) is thought to be the complete record, then there are certainly gaps in the terrestrial record.

Those trying to elucidate environmental and biotal changes through a study of continental sediments are provided in East Anglia with a wealth of sedimentary and fossil evidence. A basic question is how to organise this evidence in terms of environmental and biotal change in time, so that additional evidence can be slotted in as it comes to light. The solution which has been used is to apply the normal rules of stratigraphical classification, applying the law of superposition to lithostratigraphic and biostratigraphic units and naming type sites for particular units of environmental significance. Thus there has been suggested for East Anglia a series of stages, temperate alternating with cold, each named after a type site where there is a characteristic litho- or biostratigraphy (table 1; see also West, 1989). It should be said that the biostratigraphic units are not similar to those of earlier geological times where evolution has played a major role in the definition of units covering great lengths of time. In the Pleistocene climatic change has led to changing faunas and floras over relatively short periods of time, and it is these changes rather than evolutionary changes that are often the basis of biostratigraphy, particularly with the fossil plant assemblages.

East Anglia has been the scene of intensive Pleistocene research in the last 40 years. One result is the stratigraphic scheme in table 1. This scheme is built up from studies of many

1

Table 1. *Outline of the East Anglian Middle and Upper Pleistocene succession*

Stage name t, temperate c, cold	Notes on the characteristics of the stage, with approximate age where known, and type site or area
t Flandrian	Post-glacial forest development; named after Flandrian marine transgression in north-west Europe; began *c.* 10.0 ka BP.
c Devensian	Last cold stage; mainly periglacial in East Anglia, with ice advance to north Norfolk *c.* 18.0 ka BP; began *c.* 110–115 ka BP; type site in Staffordshire.
t Ipswichian	Last temperate stage; forest development began *c.* 127 ka BP; type site near Ipswich, Suffolk.
c Wolstonian	Mainly periglacial in East Anglia; ice advance in north-west of region; type site in west Midlands.
t Hoxnian	Temperate forest development, a period of perhaps 20 000 years; type site at Hoxne, Suffolk.
c Anglian	A major period of ice advance over much of East Anglia, with tills widely deposited; type site at Corton Cliffs, Suffolk.
t Cromerian	Temperate forest development; type site at West Runton, Norfolk (West Runton Freshwater Bed of Cromer Forest-bed).

different sites, and no one site has been found with the whole sequence. This is hardly surprising, since landscape changes associated with climatic change (e.g. ice advances, sea-level changes) preclude long depositional sequences in any one place. The scheme can therefore be criticised on the lack of superposition of the whole sequence. However, we have to classify what has been found; the result is table 1.

In spite of these developments in the study of the Pleistocene in East Anglia, there are many outstanding and important problems yet to be attacked. Are there stages additional to those listed in table 1? Some have suggested there are. What is the age of particular glacigenic sediments, such as the so-called 'cannon-shot gravels' of Norfolk? What is the extent of the ice advances into East Anglia in particular cold stages? The answers will only be provided by detailed observations in the field, and this provision is essential for the proper reconstruction of the effects of past environmental changes on our landscape and biota.

The study described in the following chapters takes advantage of the extraordinary wealth of evidence about Pleistocene environments found in a comparatively small area of central Norfolk, north of East Dereham. The time covered is from the Anglian cold stage to the present (see table 1). The field results were obtained over a long period of time while extraction of aggregates took place, using the exposures so obtained, with additional excavations and boreholes. As will be seen, the observations have led to a much better understanding of the Pleistocene in Norfolk and are indeed relevant to the wider understanding of the British Pleistocene. The results give new information on the relative age of glacial and fluviatile sediments, landscapes, biota and periglacial processes in a small area where superposition of stratigraphical units is certain.

The work underlines the importance of field observation and careful recording of sections, emphasises the importance of stratigraphy in the reconstruction of Pleistocene environments, demonstrates the variety of such environments and the processes associated with them and indicates techniques which can be used to study them.

THE AREA AND PREVIOUS RESEARCH

The investigations described here are centred on the area of Roosting Hill, near Beetley, to the north of East Dereham in central Norfolk. The area is shown in the maps (figures 1 and 2) and in the aerial photographs (figures 3 and 4). Roosting Hill was an isolated hill (see figure 3) on the western slope of the valley of the Whitewater stream but has been removed by extensive sand and gravel workings. The Whitewater joins the Blackwater stream to the north, and the Blackwater joins the River Wensum, 1.6 km to the north-east. This drainage pattern is shown on the map (figure 1). The Whitewater is a small stream, but it drains a large area to the west and north of East Dereham. It receives minor contributions from a number of springs and seepages rising on both west and east slopes of the valley near Roosting Hill. These springs are shown on the map (figure 2).

The Old Series geological map of the area (Sheet 66 NW Drift) shows extensive areas of 'Plateau Gravel', the Hungry Hill Gravels of Phillips (1976), overlying boulder clay, with valley gravels and alluvium in the valleys (Blake, 1888). The glacigenic Pleistocene deposits overlie Chalk, proved at depth in boreholes in the area. The map by Auton (1982) shows that the surface of the Chalk forms a depression below O.D. in the area of the Blackwater–Whitewater confluence, rising to over 20 m O.D. in the surrounding uplands.

The first investigations at Roosting Hill took place in 1964, when organic interglacial deposits were reported by Mr R.A.D. Markham, then at the Norwich Castle Museum. These and another nearby late Pleistocene organic deposit at Swanton Morley in the R. Wensum valley (figure 1) were subsequently studied by Phillips (1976), who concluded that the temperate stage represented was the Ipswichian. By 1964, however, Roosting Hill had gone, though it has been possible to study the original landforms from aerial photographs taken in 1946 (figures 3 and 4), before sand and gravel working started.

From 1972 onwards, further investigations took place when a number of Devensian organic deposits were exposed in the workings. Excavations were also carried out in areas stratigraphically critical for the Devensian and Flandrian. In 1982 sediments of a temperate stage older than the Ipswichian were discovered and further excavations were carried out to determine their stratigraphical position.

The area is thus extraordinarily rich in organic sediments of various ages, with the consequence that it is possible to date associated glacigenic, periglacial and fluviatile sediments at and near Roosting Hill, and to follow in some detail the morphological evolution of the Whitewater valley and so the upper part of the valley of the River Wensum. The study provides an example of a Pleistocene sequence in an upstream minor part of a river catchment, distant from the sea, where there is a fairly low local relief and less apparent landscape dissection than is seen in the lower reaches of rivers in eastern England.

The extensive record of vegetational history, though not as complete as might be found in a deep lake basin in an unglaciated area, has the advantage of providing information, through a consideration of the taphonomy of the plant fossil assemblages, about a variety of depositional environments; these can be related to various climates and geological processes, so enhancing the reconstruction of local and regional vegetation and of environmental history.

The information gained from the field studies is presented in two major ways in the illustrations. The stratigraphy is presented as detailed drawn sections, many metres in length and a few metres in depth; vertical exaggeration is present in some of the drawings. Sediment symbols used in the sections are given in figure 5. The palaeobotanical results are presented as tables detailing the composition of assemblages of macroscopic plant remains and as pollen diagrams. On the left of each pollen diagram is a summary diagram showing the percentage of tree, shrub and herb pollen. All the percentage values are based on the total of all pollen of land

3

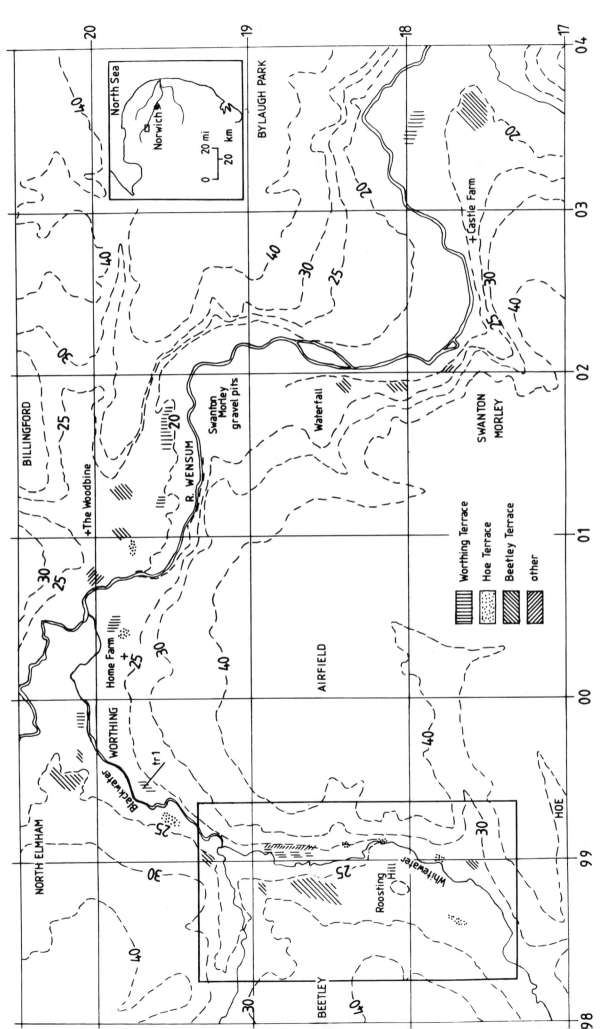

Figure 1. Location of the study area at Roosting Hill. Inset right, East Anglia; inset left, study area.

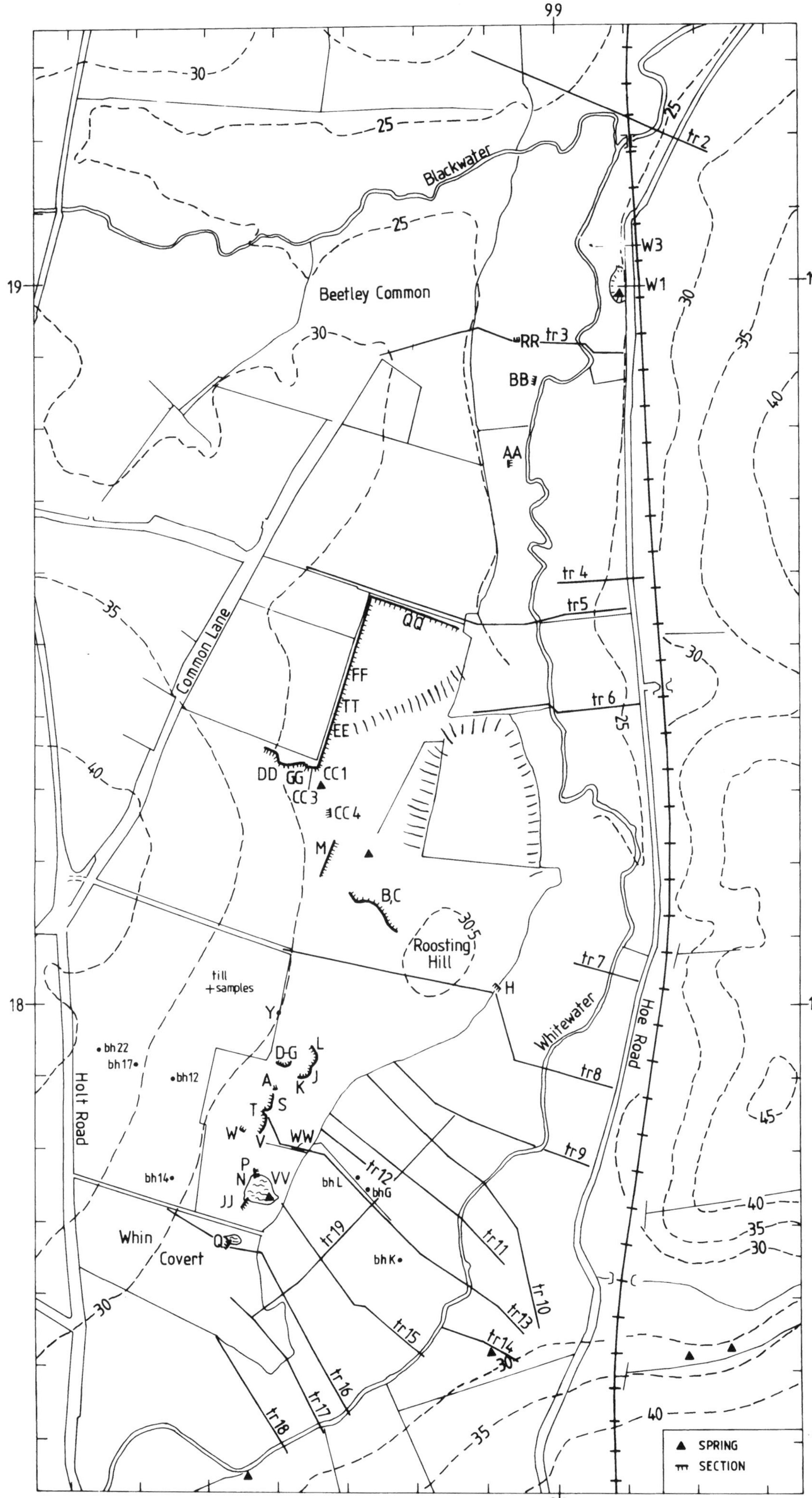

99

30

25

Blackwater

25

19

Beetley Common

W3

W1

RR tr3

BB

AA

tr 4

Qq

tr5

FF

TT

EE

tr 6

DD GG CC1

CC3

CC4

M

B,C

Roosting
Hill

30·5

35

Common Lane

till
+ samples

H

tr 7

18

Y

bh 22

bh 17

D-G L

bh 12

A K J

T S

W V WW

P

N VV

bh L bh G

JJ

Q

Whin

Covert

tr 19

tr 12

tr11

bh 14

bh K

tr 10

tr13

tr 15

tr14

tr 16

tr 17

tr 18

Whitewater

Holt Road

Hoe Road

30

25

40

45

40

35

30

35

40

99

19

18

SPRING

SECTION

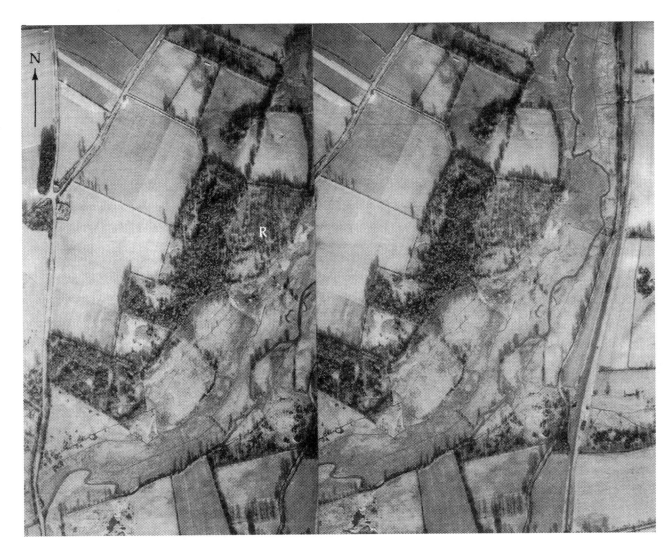

Figure 3. Stereo pair of aerial photographs of the southern part of the study area, taken in January 1946, before sand and gravel extraction started. R, Roosting Hill.
Crown copyright/RAF photograph.

Figure 4. Aerial photograph of the northern part of the study area, taken in January 1946, before sand and gravel extraction started. B, Beetley Terrace; Bw, Blackwater; H, area of hollows; R, Roosting Hill; Wh, Whitewater; Wl, hollow Wl.
Crown copyright/RAF photograph.

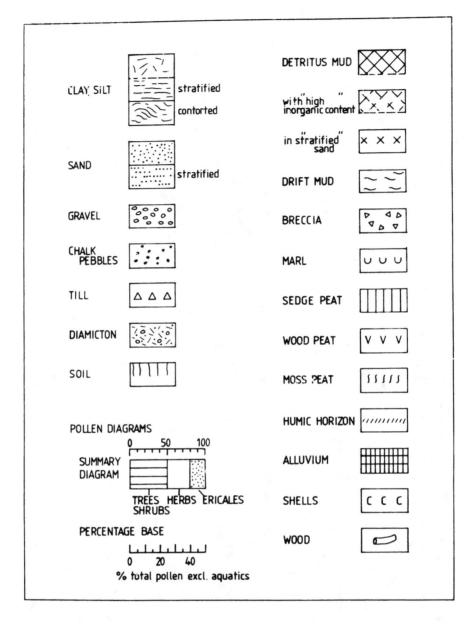

Figure 5. Key to sediment symbols used in sections and pollen diagrams.

plants counted in each analysis, with Cyperaceae pollen included in this total. Percentages of pollen of aquatic plants and of spores of lower plants are shown on the right of the diagrams, together with counts of derived (secondary) pollen and spores. The crosses indicate presence of a taxon, with abundance reflected in the number of crosses.

Appendices I and II provide respectively notes on certain techniques used and notes on certain general aspects of the palaeobotany. Plant nomenclature follows *Flora Europaea* (Tutin *et al.*, 1964–80).

OUTLINE OF STRATIGRAPHY AND LANDSCAPE HISTORY

Since the detailed stratigraphy to be described is complex, it is necessary first to give an outline of the major lithostratigraphical units distinguished, listed in table 2. The stratigraphical position of the various sections is shown in figure 2. The relationship of the units in the investigated area is shown in the north-south section, figure 6, and the general location of sediments of the different stages in figure 7.

Bed A   Outwash gravels, till. A thick complex of glacigenic sediments overlies Chalk in the area of Roosting Hill. These sediments have been described by Phillips (1976) and by Auton (1982), who gives a map showing that they lie in a local depression in the Chalk surface which extends below O.D. The thickness reached is over 29 m, and the sediments include the so-called cannon-shot gravels or Hungry Hill Gravels of Phillips (1976) and till (chalky boulder clay). Exposures of the latter in the gravel pit south-west of Roosting Hill (figures 2 and 7) have been studied by Ehlers, Gibbard & Whiteman (1987), who record a lower dark grey till overlain by a chalk-rich pale till under several metres of Hungry Hill Gravels. Fabric analyses showed ice movement from the south-west for the former and from the north-west for the latter till. The bedding of the overlying gravels also clearly indicates derivation from the north-west. Although there is much variation in the lithology of this unit, there is no evidence that more than one cold stage is represented. The till extends eastwards, being recorded at depths of 5–10 m in the Whitewater valley.

Bed B   Limnic sediments: silts and marls. Overlying Bed A gravels and sands in sections DD and TT (figures 8, 9 and 10) are limnic sediments. The stratigraphy of the limnic sediments in DD and TT differs, implying two basins of deposition within a short distance of each other. The basal limnic sediments in both basins are late-glacial in vegetational character, indicating an irregular late-glacial topography left by ice retreat. The younger limnic sediments show the development of temperate forest. The maximum thickness of these limnic sediments is 2 m.

Bed C   Gravels forming a terrace, named the Beetley Terrace. These gravels overlie the limnic sediments of Bed B in section TT (figure 9) and extend north through the contiguous sections FF and QQ where they form a low terrace fronting the west margin of the floodplain of the Whitewater stream (figures 7 and 13). In section DD (figure 8) they overlie the limnic sediments of Bed B and, to the west in this section, thin out over the sands and gravels of Bed A. The sediments are ill-sorted angular flint gravels with subordinate seams of sand and silt, reaching a maximum thickness seen of 3 m; they are considered to be part of a cold stage aggradation.

Bed D   Fluviatile sands and associated organic sediments. In 1964 sections B and C (figures 16 and 17) showed a sequence of two beds of sandy coarse mud separated by fluviatile sands with fine flint gravel, totalling 1.5 m in thickness. The flora and fauna of these sections indicate temperate conditions. In section DD (figure 8) a similar sandy coarse mud, with a similar pollen

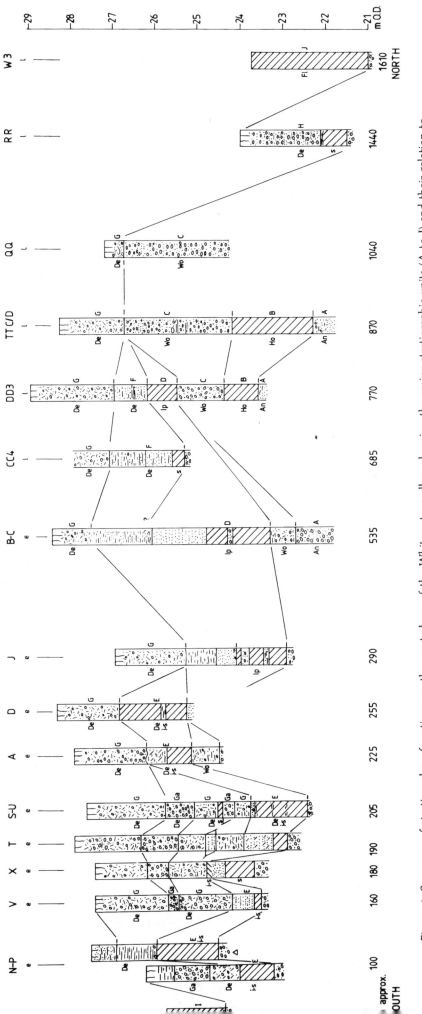

Figure 6. Summary of stratigraphy of sections on the west slope of the Whitewater valley, showing the major stratigraphic units (A to J) and their relation to stages (An, Anglian; Ho, Hoxnian; Wo, Wolstonian; Ip, Ipswichian; De, Devensian; Fl, Flandrian; i-s, interstadial; s, stadial).

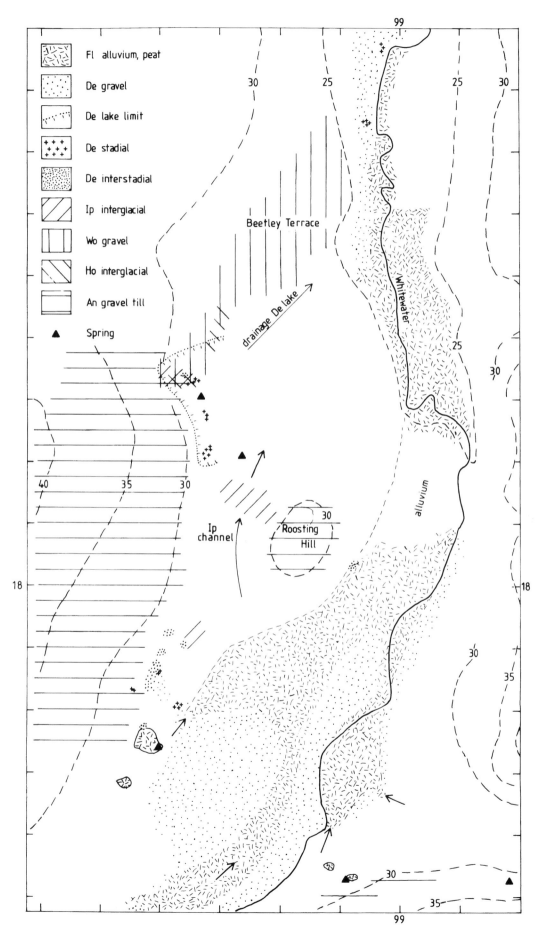

Figure 7. Location of sediments of the several stages identified. Springs and drainage directions (arrows) are shown.

Table 2. *Lithostratigraphical units*

Bed	Lithology	Approximate maximum thickness recorded	Stage
J	Alluvium: clays, peats	3 m	Flandrian
I	Limnic sediments in hollows: organic sediments, marls	3 m	Flandrian, Late Devensian
H	Floodplain gravels, including sands, silts, organic sediments	>4 m	Devensian
G	Solifluction diamictons and associated fluviatile sands and gravels (Ga)	2 m	Devensian
F	Limnic sediments (widespread): sands, silts, clays, organic sediments	2 m	Devensian
E	Limnic and telmatic organic sediments in depressions	1 m	Devensian (interstadial)
D	Fluviatile sands and associated organic sediments	1.5 m	Ipswichian
C	Gravels forming Beetley Terrace	3 m	Wolstonian
B	Limnic sediments: silts, marls	2 m	Anglian late-glacial, Hoxnian
A	Outwash gravels, till Chalk	29 m	Anglian

flora at its base to that of section B, is seen to overlie the gravels of Bed C. The sediments of bed D are thus associated with fluviatile temperate conditions post-dating the aggradation of the gravels of Bed C. Further organic sediments ascribed to the same temperate stage were seen in section J–K–L (figure 20).

Bed E Limnic and telmatic organic sediments, including peat, of isolated sequences. The sediments were rarely over a metre in thickness and occupied depressions in subjacent sands and gravels. They occurred in section GG (figure 30), section A (figure 22), section D–E–F–G (figure 22), section K (figure 21), sections R–S–T–U–X–Z–V (figures 24 and 28), and section N–P (figure 31). The palaeobotany of these depression fillings is varied, with assemblages ranging from boreal forest to cold stage herbaceous vegetation.

Bed F Limnic sediments (widespread): sands, silts, clays, organic sediments. Overlying the organic sediments at section GG of Bed E is a sequence of limnic sediments, which are widespread between sections DD, C and M (figures 8 and 30). At the base is a thin seam of coarse mud (e.g. sections CC, M), which is overlain by first sand and then silt and clay, the whole reaching a maximum thickness of nearly 2 m. The sequence indicates the establishment of a rising water level and lacustrine conditions in a tributary of the valley at this time (figure 7).

Bed G Solifluction diamictons and associated fluviatile sands and gravels. On the western slope of the Whitewater valley, a sheet of solifluction sediments, in places distorted by festooning, mantles the gravels of Bed A upslope and the sands and gravels of Bed C and the silty clays of Bed F downslope. Sections DD, FF and QQ (figures 8, 9 and 13) show this superposition. The maximum thickness of the diamicton, composed of silt, sand and gravel, is about 2 m. Section QQ (figure 13) shows the solifluction sheet thinning towards the eastern edge of the terrace associated with Bed C. Further south, in sections R–V (figures 24 and 28), solifluction diamictons overlie Bed E sediments and are associated with fluviatile sands and gravels, separated as Bed Ga in figure 6.

Bed H Floodplain gravels, including sands, silts and organic sediments. These fluviatile sediments, predominantly flint gravel, lie under the alluvium of the Whitewater valley and reach a thickness of over 4 m; they overlie Anglian till. Their relation to the terrace to the west is shown in section QQ (figure 13). The gravels contain channels filled with organic sediment and silty horizons with drift mud, as seen in sections H, AA, RR and BB (figure 13). Radiocarbon dating indicates this unit is Late Devensian in age in its upper part.

Bed I Limnic sediments in hollows: organic sediments, marls. Small hollows at the floodplain margin and at higher levels contain Devensian late-glacial and Flandrian sediments to a depth of 2–3 m. These appear to be related to springs and associated subsidence.

Bed J Alluvium: clays, peats. These sediments comprise an upper alluvial clay and a lower humified peat, of a maximum depth about 3 m, filling channels of the present Whitewater valley.

Before considering the detail of the succession at Roosting Hill, it will be helpful to outline the probable correlation of these units to the standard East Anglian succession (see table 1). The pollen diagrams allow correlation of the temperate stage sediments of Beds B and D to particular temperate stages whose position in the Pleistocene succession in East Anglia is well-established. The pollen diagrams from Bed B sediments, as will be discussed later, are similar to Anglian late-glacial and early Hoxnian diagrams known in Norfolk and elsewhere in East Anglia. Pollen diagrams from Bed D have been correlated by Phillips (1976) with diagrams from the Ipswichian (temperate) Stage, such as those from Swanton Morley, 3 km to the east in the R. Wensum valley. If these correlations of the temperate stages at Roosting Hill are correct, we may place the cold stage sediments in the Anglian (Bed A), Wolstonian (Bed C) and Devensian (Beds E, F, G, H, I). These correlations do not contradict any present understanding of the relative age of cold stage sediments in Norfolk. They are summarised in table 2.

The outline of the succession at Roosting Hill given above now allows an outline of landscape history in the area to be reconstructed, as a preliminary to the more detailed accounts of environmental history which follow.

The area was subjected to ice cover during the Anglian (cold) Stage, during which tills and outwash gravels were deposited. On the retreat of the ice an irregular topography was exposed with separate basins in which limnic sedimentation started in the late-glacial and continued into the subsequent temperate stage (Hoxnian). In the cold stage which followed (Wolstonian) the landscape must have been much reduced in relief, with a drainage system established similar to that at present, and an aggradation (Bed C) forming a terrace. Fluviatile and organic sediments of the subsequent temperate stage (Ipswichian) later formed in the same valley, reaggrading after incision of the terrace.

Later than the deposition of the Ipswichian sediments, a series of organic deposits were formed in localised hollows, of early Devensian age. These were subsequently covered by sediments associated with limnic conditions and a rising water level in the valley, first organic muds, later sands and silty clays. The possible origin of this water-level change is discussed later. These limnic sediments were subsequently covered by a solifluction sheet, smoothing the landscape to a state which is seen today. The production of a solifluction sheet over permeable sands and gravels indicates permafrost in the latter part of the Devensian. Incision of the valley was followed by aggradation of the floodplain gravels which underlie the Flandrian alluvium and peat. Flandrian sediments are also found in depressions associated with springs and/or subsidence.

CHAPTER TWO

The Anglian (cold) Stage and the Hoxnian (temperate) Stage

As already described, the retreat of the Anglian ice left an irregular topography, and two lake basins about 80 m apart associated with this landscape have been investigated. The lake sediments form Bed B of the succession. The age of the local 'cannon-shot' (Hungry Hill) gravels which underlie the lake sediments has always been a matter of argument. The question has been whether the great till sheet which was spread across East Anglia during the Anglian (cold) Stage, the chalky boulder clay, belongs to the same ice advance as the 'cannon-shot' gravels. At Beetley this till occurs closely associated with the gravels, and it is possible to demonstrate clearly that till and gravels are a result of the same pre-Hoxnian ice advance, since the lake sediments show a transition from an open landscape at the end of the cold stage to a forested landscape in the Hoxnian. The stratigraphy of the lake basins is first examined and then the vegetational history described and analysed.

STRATIGRAPHY

Sediments of the two lakes are seen in sections DD (figure 8) and TT (figures 9 and 10), the positions of which are shown in figure 2.

Section DD At the eastern end of this section (1–5 m) limnic sediments were found resting on sands and gravels of Bed B and overlain by the flint gravels of Bed C. The sediments, maximum thickness c. 1.3 m, consisted of brown silty clay at the base, then pale marl, then khaki clay-mud, as shown in the stratigraphy column of the pollen diagram DDB (figure 11) from samples at the 1-m point of the section DD. To the west the lake sediments are cut out by the overlying gravel.

Section TT This section, in the ditch running north from near section DD, is in the central part of the long section running north to the northern margin of the old gravel pit. The whole section is given in figure 9, which shows the stratigraphic relation of the lake sediments to Beds A and C. Figure 10 shows an enlarged version of the section, and also the relation to the section DD Bed B basin (at point DD 1 m). The stratigraphy of the TT basin is different from that of DD. It is more complex and the marl occupies a different position in the sequence. As will be seen later, the pollen diagrams also differ, though sharing a common pattern, indicating taphonomic differences of the pollen assemblages in each basin.

The basal sediment of the TT lake sequence is a grey calcareous silt, the silt probably derived from the local silt-rich till (see figure 41). At TTC (111 m) this is nearly 2 m thick and is overlain conformably by a pale marl 0.2 m thick. Both the grey silt and the marl have low organic content (table 3) and appear to be sediments of an open system lake. To the south, at 91.5–94 m, the grey silt is separated from the marl by a wedge of sediments. This wedge is itself divided into two parts, a lower grey silt more organic at the base, and an upper brown

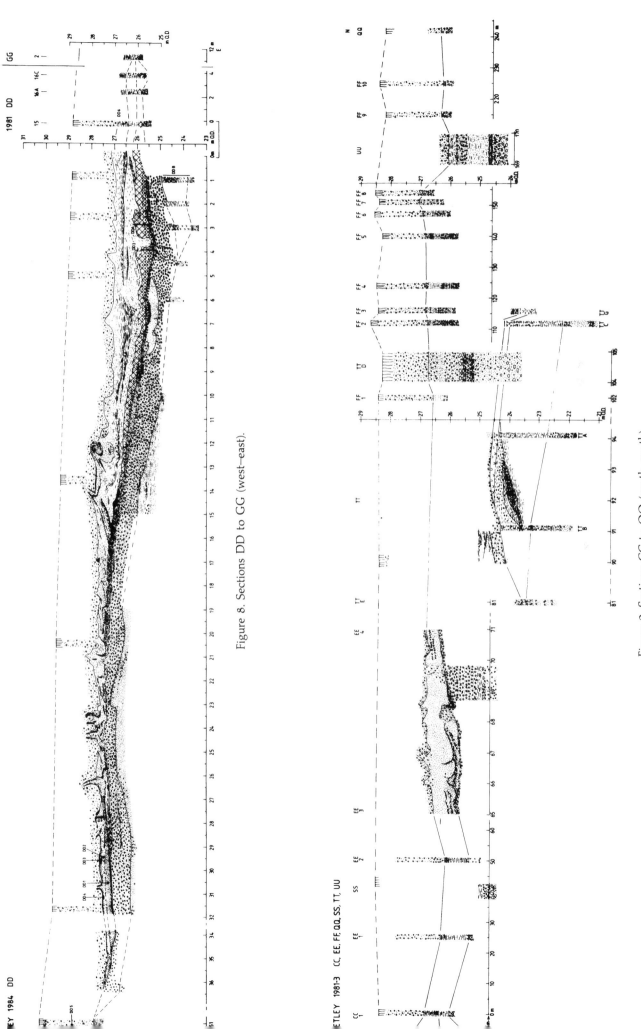

Figure 8. Sections DD to GG (west–east).

Figure 9. Sections CC to QQ (south–north).

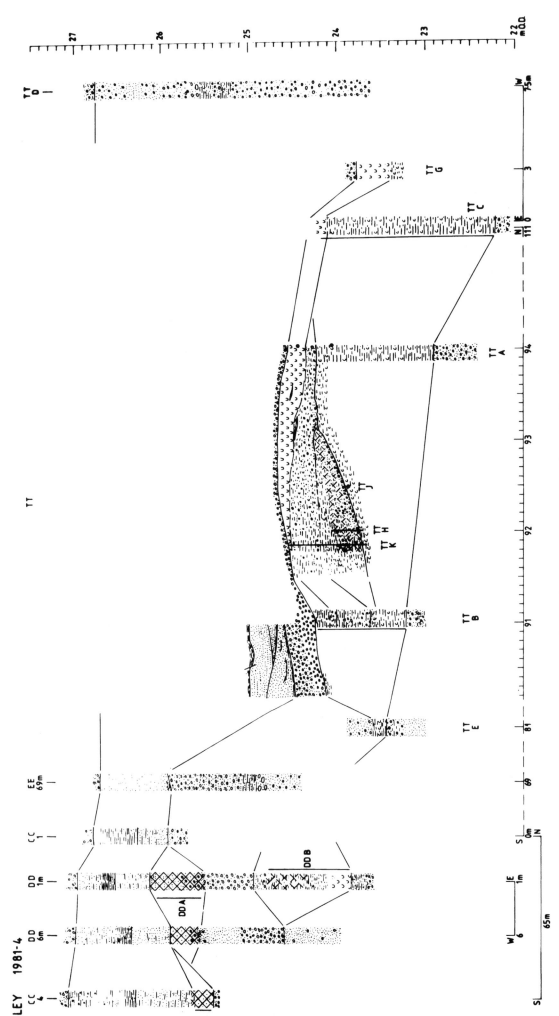

Figure 10. Correlation of sections CC, DD, EE and TT (south–north).

Table 3. *Percentage dry weight loss-on-ignition (550 °C) of selected samples*

	Site	Depth (cm)	% loss		Site	Depth (cm)	% loss
Early Flandrian	Q3	95	16.5	Ipswichian	B	150	4.5
Devensian					C	10	7.0
late-glacial					C	40	6.7
interstadial	H	4	49.7		C	70	11.7
stadial	AA	30	18.9		C	75	16.5
	AA	79	14.9		C	78	47.2
	RR	30	10.2		DD3	6	53.6
	M	10	64.9		DD3	30	33.0
	CC4		39.3		DD3	50	5.0
	R	10	9.1		DD3	60	7.7
	L		63.6		DDA	11	33.0
Early Devensian					DDA	15	73.9
interstadial	UA	10	3.8		DDA	105	7.0
	UA	35	4.8		J	0	10.0
	A	5	6.3		J	40	5.0
	A	30	20.3		J	75	3.9
	A	50	37.5	Hoxnian	J	85	5.6
	A	60	16.2		TT		
	T	10	13.1		marl		2.4
	T	25	36.5		TTC	30	2.9
	T	30	68.9				
	T	35	47.5				
	T	40	5.9				
	S	20	12.1				
	S	70	10.0				
	S	75	9.6				
	S	80	6.1				
	U	0	10.4				
	K	10	5.6				
	GG	30	4.5				
	GG	10	78.2				
	GG	5	77.2				

silty clay with small flint and chalk pebbles and thin seams of sand. The detail of this wedge is seen in the stratigraphic column of the TTK pollen diagram (figure 12). The wedge is clearly overlain by the pale marl. It is evidently a result of landslip into the lake basin. The pollen diagrams from TTC, TTA, TTH, TTJ and TTK described in the next section enable an analysis of the landslip process. The lower part of the slip appears to have taken place by slip with the preservation of the stratigraphy of an earlier part of the lake filling. The upper part of the slip, with its more inorganic unsorted and weathered nature, appears to be a slump into the lake extending beyond the lower part, but not reaching TTC. The landslip evidently caused a change of sedimentation in the lake, with a change from grey silt to marl. This implies that there was a change from an open system lake with an inorganic input to a less open type of lake in which marl formation was favoured. The marl contains *Chara* oospores but not in any abundance, and is not a *Chara*-marl. It has a low organic content (table 3), and its origin may lie in an increased supply of local carbonate-rich spring waters, following the landslip. There is an interesting similarity of this sedimentary 'interruption' in the early part of the temperate stage with those seen at Marks Tey (Turner, 1970) and Fishers Green (Gibbard & Aalto, 1977). The comparison will be discussed when the pollen diagrams from TT have been described. Mollusc assemblages from the grey silt and marl adjacent to section TTC are described by R.C. Preece in Appendix III.

A further complication of the TT section is seen in the TTB section (91 m). Here the stratigraphy, shown in the pollen diagram in figure 12, is more irregular than in the basal part of TTC, with evidence of slip or disturbance at 60 cm depth. At a higher level, above a horizon at 16–19 cm with chalk pebbles and shell fragments, a sudden change in the pollen content implies an unconformity with incision of the earlier sediments and deposition of grey and brown clays later in the temperate stage. This change is likely to be associated with a further change in the local drainage system, with the replacement of limnic conditions by the aggradation in a channel of inorganic alluvial-type sediments.

VEGETATIONAL HISTORY

Pollen diagrams have been constructed from a number of sections: DDB in the DD section at 1 m; TTA, 94 m; TTB, 91 m; TTC, 111 m; TTH, 92 m; TTJ, 92.5 m; TTK, 91.8 m. The lower parts of DDB, TTB and TTC are from boreholes, the remainder from open section. As with all the pollen diagrams, subdivision is made into pollen assemblage biozones (p.a.b.), named after important pollen taxa of the biozones. Correlation of the pollen assemblage biozones with regional biozones is shown in table 4.

DDB (figure 11) This diagram shows three pollen assemblage biozones:

    c. 10–50 cm. *Quercus–Betula* p.a.b. Higher frequencies of *Quercus* and lower frequencies of *Betula*.

    b. 50–100 cm. *Betula* p.a.b. high frequencies of *Betula*, with low frequencies of *Pinus*, *Ulmus*, *Quercus* and Gramineae.

    a 115 cm. *Hippophae* p.a.b. High frequencies of *Hippophae*, with low frequencies of *Betula* and *Salix*.

This sequence shows the replacement of late-glacial *Hippophae* communities with *Betula* and *Salix* by *Betula* woodland with low frequencies of thermophilous trees, including *Ulmus* and *Quercus*. *Quercus* then increases its representation in the uppermost biozone. This vegetation history is typical of the transition from a cold stage to a temperate stage in East Anglia and resembles closely late Anglian to early Hoxnian successions seen elsewhere in East Anglia (West, 1980), in particular the high frequencies of *Hippophae* in the late-glacial, not yet known from late-glacials of other cold stages, and its later replacement, at first by *Betula* and Gramineae (as at Hoxne; West, 1956), then by *Betula*, and the subsequent development of forest with *Quercus*. The three biozones of DDB can then be related to the regional biozones l An, Ho I and Ho IIa.

TTC core (figure 11) This TT diagram is taken first, since it shows an undisturbed sedimentary sequence through limnic sediments overlying the sands and gravels of Bed A. Three pollen assemblage biozones are present in the TTC core:

Table 4. *Anglian late-glacial and Hoxnian pollen assemblage biozones*

Substage	Biozones of pollen diagrams				Pollen assemblage biozone
	DDB	TTC core	TTB	TTH, TTK	
Ho IIb			e		*Alnus–Quercus–Betula–Pinus*
Ho IIa	c				*Quercus–Betula*
Ho I	b	c	d	< landslip	*Betula–Pinus*
		b			*Betula*
l An	a	a	b		*Hippophae*
			a		*Salix*–Gramineae

c. 0–110 cm. *Betula–Pinus* p.a.b. Higher frequencies of *Pinus* pollen are present. *Ulmus*, *Quercus* and *Juniperus* also present.

b. 110–200 cm. *Betula* p.a.b. High frequencies of *Betula* pollen, with lesser frequencies of *Pinus*, *Salix*, *Juniperus* and Gramineae.

a. 205 cm. *Hippophae* p.a.b. High frequencies of *Hippophae* pollen, with lesser frequencies of *Betula*, *Salix*, *Pinus* and Gramineae.

This succession again shows the transition from late-glacial vegetation to the early part of a temperate stage. *Hippophae* communities are again present at the base, followed by an expansion of *Betula* woodland, with *Pinus* becoming more important later and *Ulmus* and *Quercus* only represented at very low frequencies. The earliest biozone can be related to the regional biozone l An, and the two later biozones to Ho I.

Comparison of DD and TTC core   The pollen diagrams from these sites differ in a number of ways. In the DDB *Betula* biozone *Pinus* pollen is much less frequent, *Juniperus* pollen less frequent and *Quercus* more frequent, and the diagram also extends upwards into a biozone with high *Quercus* frequencies, so representing a longer period of time than the TTC core diagram. These differences are best explained by interpreting the diagrams as representing two distinct basins of deposition, with different sediment sequences. The TTC sediment has a high inorganic content and the higher frequencies of *Pinus*, *Juniperus* and Gramineae pollen can be related to an open lake system deriving pollen from a large catchment. On the other hand, DDB may be the site of a closed system lake, with a lower sedimentation rate and with local pollen much better represented, leading to lower frequencies of these three taxa.

TTH, TTA, TTJ (figure 12)   When the TTH section (figure 10) was first seen and analysed for pollen, and before the section was extended north and south to relate it to the other sections, the pollen analyses with high frequencies of *Hippophae* pollen were unexpected, since they overlay sediments with high *Betula* pollen frequencies of the early part of a temperate stage. On excavation, however, the section seen in figure 10, 91–94 m, was exposed, showing that the TTH section was part of a wedge of sediment intruding into the lake sequence. The TTH pollen diagram, 25 cm in depth, is from the lower part of this wedge. It shows a rise of *Hippophae* pollen to a high level, followed by a fall; *Betula* pollen responds in the reverse way. Gramineae pollen is higher at the base, *Pinus* pollen higher at the top. At the top of the underlying sediment, the grey calcareous silt, analyses at TTJ and TTA 40 cm belong to the *Betula–Pinus* p.a.b. of TTC. The conclusion is that the slab of sediment TTH, late Anglian in part, has slid down into the lake basin during Ho I from a more marginal part of the lake. In

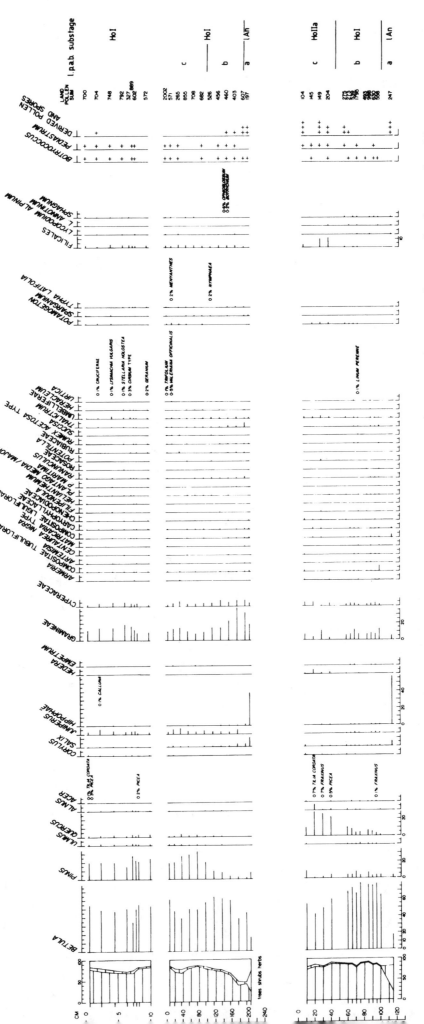

Figure 11. Anglian late-glacial and Hoxnian pollen diagrams, sections TTC and DDB. On the left of the sediment column are indications of sediment colour: bl, black; br, brown; dg, dark grey; f, fawn; g, grey; p, pale. Percentages expressed as percentage of total pollen excluding aquatics.

view of the mode of deposition of TTH a detailed interpretation of the pollen diagram is not possible, apart from saying that the lower part is probably all of late-glacial origin while the upper part includes Ho I sediment. The higher than elsewhere frequencies of Gramineae and Cyperaceae pollen are probably related to the originally more shallow water and marginal position of the TTH late-glacial sediment.

TTK (figure 12)   A further monolith was taken through the wedge of sediment at TTK when its possible landslip origin became clearer as a result of excavation. The basal sample belongs to the underlying grey calcareous silt and the top sample to the pale marl overlying the wedge sediments. From 35 to 75 cm is grey calcareous silty clay, more organic near the base, with a seam of sand, flint pebbles, and chalk pebbles at 60 cm. This part of the diagram, at the base of the landslip, is similar to the TTH diagram, showing a peak of *Hippophae* pollen. There is a very sharp boundary with the overlying sediment at 35 cm, marked by a line of small flint pebbles. This overlying sediment, from 5 to 35 cm, is red and brown silty clay with occasional sand seams and small flint and chalk pebbles. The more weathered appearance of these sediments suggests a part origin in weathered sediments. The *Pinus* pollen percentages are consistently higher in the wedge part of the TTK diagram than in the TTC core diagram, possibly because the source Ho I sediments had a higher proportion of *Pinus* pollen or because of a source of pollen in a local soil bank. In the uppermost sample in the pale marl, *Betula* and *Pinus* percentages return to the level seen in the *Betula–Pinus* biozone (Ho I) in TTC. Thus the whole sequence of TTK must have been deposited in Ho I. The upper weathered part of the landslip can be seen in the TT section to extend north beyond the lower part by over a metre, giving the impression that the wedge of sediments was emplaced in two steps, an early slide with original stratigraphy preserved to some extent and a later slump of weathered more heterogeneous sediment.

TTC monolith, TTA (figures 11 and 12)   TTA is at the north end of the landslip wedge where it is only 15 cm thick. The analyses show, as in TTK, the continuance of Ho I over the landslip period. The TTC monolith is beyond the limit of the wedge; the samples here were taken at very close intervals across the boundary between the grey calcareous silty clay and the pale marl. Since the sediments here appear conformable, it was hoped that a close-sampled diagram would reveal any vegetational changes associated with the landslip. The transition is marked by a rise in Gramineae pollen shown in the three analyses at 0.5-cm intervals, but apart from this there is no great change in the pollen deposition or in the variety of non-tree pollen, such as might be expected from an increase in herbaceous vegetation following a massive landslip. This would suggest a major regional component in the pollen input, perhaps derived from the catchment in the open-system lake.

TTB core and monolith (figure 12)   This pollen diagram adds further detail of the Anglian late-glacial and provides evidence for a later pollen assemblage biozone than is seen in the other diagrams. The following pollen assemblage biozones are distinguished:

>   e. 0–16 cm. *Alnus–Quercus–Betula–Pinus* p.a.b. *Tilia* and *Corylus* are also present in significant frequency. The abrupt change in pollen frequencies associated with a horizon of small chalk clasts at the base of this biozone indicates an unconformity.

>   d. 19–65 cm. *Betula–Pinus* p.a.b. Similar to the TTC core biozone of the same title; the sediment is brecciated in part and contains small flint and chalk clasts, so there may be reworking.

>   c. 65–100 cm. *Betula–Pinus–Hippophae* p.a.b. This assemblage covers a period of irregular sedimentation and variable pollen content. At 65 cm a horizon with high

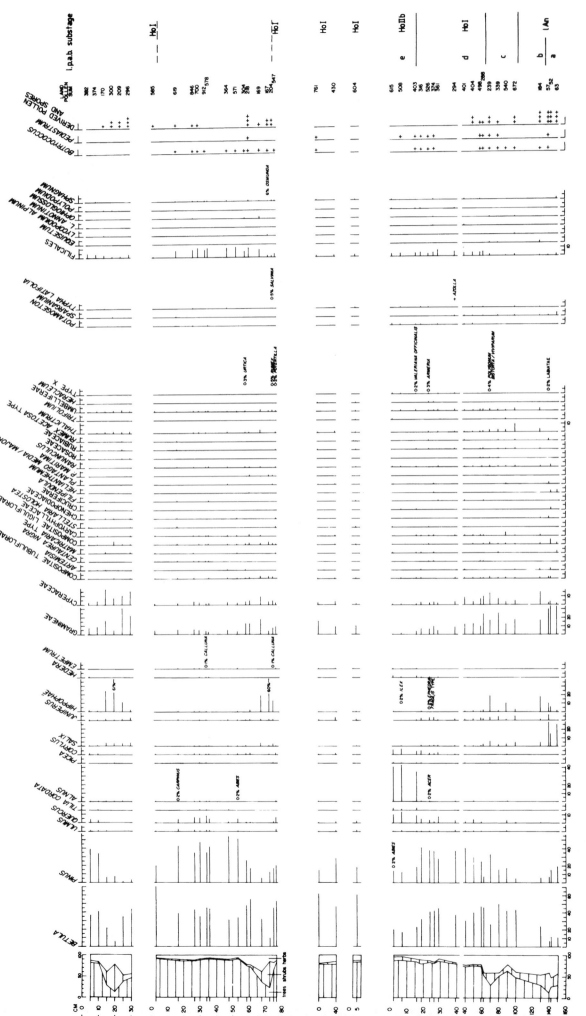

Figure 12. Anglian late-glacial and Hoxnian pollen diagrams, sections TTA, TTB, TTH, TTJ, TTK. Sediment colour indications as in figure 11. Percentages expressed as percentages of total pollen excluding aquatics.

Hippophae frequency resembles the lithology and pollen content of the basal sediments of the landslip at TTH and TTK.

b. 130 cm. *Hippophae* p.a.b. *Hippophae, Betula,* and Gramineae are the prominent taxa at this horizon.

a. 140–150 cm. *Salix*–Gramineae p.a.b. *Salix* and Gramineae are the predominant pollen taxa, with *Betula, Pinus, Juniperus, Hippophae* and a variety of herbs also present.

The sequence shows an early late-glacial assemblage with *Salix* and Gramineae prior to the *Hippophae* community which follows. The middle part is confused with evidence for reworking and landslip. The uppermost biozone shows the presence of mixed oak forest with *Ulmus, Tilia, Quercus, Alnus* and *Corylus*. This biozone is correlated with the regional biozone Ho IIb.

The origin of landslip in Ho I   During Ho I instability must have been induced at the margins of the TT lake. Such instability could be caused by local artesian water pressures, by collapse associated with solution of the Chalk below, by earth tremor, or by fluctuation of the water table in a steep-sided basin. Slumping of lake marginal sediments following lowering of water level has been recorded in lakes (e.g. Reis, 1931), but there is no evidence of a break in succession or reworking in the limnic sequence beyond the area with the landslip (TTC section).

Slumping in Ho I is also recorded in the deep lake sequence at Marks Tey, Essex (Turner, 1970). Changes of water level are discussed by Gibbard & Aalto (1977) as contributing to breaks in limnic sediments in the Hoxnian. They recorded such breaks, indicating a drop in water level, in Ho I at a number of Hoxnian sites in Hertfordshire. The similarity in timing of slumping at Roosting Hill and Marks Tey is remarkable and could indicate a common cause. If so, the cause was a regional effect and such a regional effect could include an earth tremor.

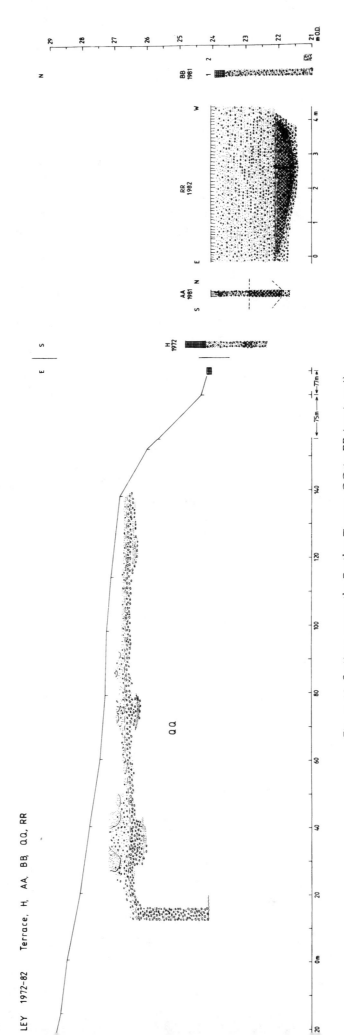

Figure 13. Sections across the Beetley Terrace, QQ to BB (west–east).

CHAPTER THREE

The Wolstonian (cold) Stage

Evidence for a cold stage climate in East Anglia directly following the Hoxnian temperate stage has been found at a number of sites. Likewise, evidence for a cold stage climate before the later Ipswichian (temperate) Stage has also been recorded at a number of sites. The cold stage thus identified between the Hoxnian and Ipswichian, termed the Wolstonian after a site in the west Midlands, has been the subject of much discussion. The problems are of two kinds. First, was there an ice advance into East Anglia during this cold stage, and, if so, how extensive was it? Secondly, how complex is the cold stage, and is the cold stage climate interrupted by interstadial or even temperate conditions? The significance of the evidence at Beetley is that it shows the presence of fluviatile gravels forming a terrace sandwiched between Hoxnian and Ipswichian sediments. The gravels must be allocated to the Wolstonian (cold) Stage. There is no evidence for an ice advance into the area between the Hoxnian and Ipswichian; rather it is a period of periglacial conditions, with aggradation of a terrace and the formation of solifluction diamicton after incision into this terrace. The Wolstonian sequence at Beetley thus appears to be relatively simple, though this simplicity is certainly a result of the lack of preservation of the whole Wolstonian record, which elsewhere is far more complex.

Three long sections show the relationship of Bed C gravels to temperate stage sediments and a terrace of the Whitewater stream, here named the Beetley Terrace:

Section QQ (figure 13)   This east–west section at the northern end of the gravel pit cuts across the Beetley Terrace. It shows coarsely stratified angular flint gravels (Bed C), with subordinate lenses of stratified sands, overlain by a solifluction diamicton affected by involutions (Bed G).

Section EE to QQ (figure 9)   The gravels of section QQ can be seen along the whole section southwards to EE, underlying the solifluction diamicton. At TT they overlie the Anglian late-glacial and Hoxnian limnic sediments already described. At EE 3 to EE 4 the sands and gravels begin to be cut out, the infilling sediment being sands of Bed F.

Section DD (figures 8, 14 and 15)   This east–west section shows a bed of flint gravels younger than the Bed A sands and gravels. To the west these gravels wedge out between Bed A and the solifluction diamicton of Bed G. Towards the east their level descends and they are overlain first by the sands of Bed F and then by the temperate organic sediments of Bed D. At the east end of the section they are underlain by temperate limnic sediments of Bed B, as in section TT.

The stratigraphical evidence clearly places the gravels of Bed C, on which the Beetley Terrace is developed, in a period between temperate stages. Their structure indicates cold climate deposition, and their disposition over Hoxnian deposits suggests a great change in landscape was effected before their deposition.

Figure 14. Section DD (figure 8), 18–27 m. Anglian indurated red sands at the base (A), pale Wolstonian fluviatile gravels (C), disturbed Ipswichian organic sediments (D) and Devensian sands F, and Devensian solifluction diamicton G.

Figure 15. Section DD (figure 8), near 23 m. Lettering as in figure 14.

The surface of the gravels of Bed C in the QQ terrace section lies near 27 m O.D. and reaches 27.5 m O.D. at the western end of section DD. Assuming that these levels are likely to be the height of the aggradation, they can be used as a basis for relating the Beetley Terrace at Roosting Hill to terraces elsewhere in the area and downstream in the R. Wensum valley. This will be done in a later section.

A further sediment which can be ascribed to this cold stage is the solifluction diamicton lining the channel in which Ipswichian organic sediments lie in section B–C (figures 16 and 17). This channel must post-date the gravel aggradation which led to the formation of the Beetley Terrace.

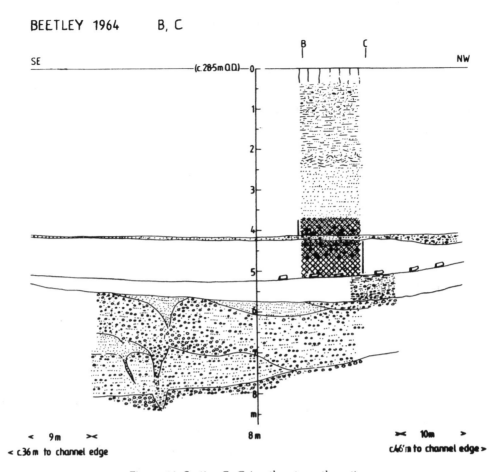

Figure 16. Section B–C (southeast–northwest).

Figure 17. Section B, C (figure 16). Ipswichian organic sediments (D) overlying Anglian or Wolstonian gravels, with a cast of a thermal contraction crack intervening.

CHAPTER FOUR

The Ipswichian (temperate) Stage

Organic sediments with temperate plant remains and rich vertebrate faunas have been found at several sites in East Anglia. That at Bobbitshole, near Ipswich (see West, 1980), is the type site, hence Ipswichian for the name of the temperate stage. The organic sediments usually lie in present-day valleys and are associated with fluviatile gravels of the preceding and following cold stages. The forest history of the stage in East Anglia is distinct from that of the preceding temperate stage, the Hoxnian. A remarkable characteristic is the presence of a rich vertebrate fauna, sometimes associated with relatively tree-less and herb vegetation. The record of this temperate stage at Beetley is confined to early and later parts of the stage, with clear super-position of the sediments on gravels ascribed to the Wolstonian which themselves overlie Hoxnian sediments. Such a clear relation between Ipswichian and Hoxnian sediments has not been found elsewhere. In addition, Beetley gives an excellent example of a site with a rich temperate vertebrate fauna associated with herb vegetation, discussed further below. The Ipswichian sections are complex, with a varying palynology. Their stratigraphy and palaeobotany are now to be described, together with an account of the palynology associated with the vertebrate fauna.

Phillips (1976) described Ipswichian organic sediments recorded in section B–C in 1964. Further studies have now been made of organic sediments, ascribed to the same temperate stage, in sections DD (figure 8) and J–K–L (figure 20) and of sediment from further bones found in the temperate stage sediments of section B–C. Table 5 summarises the biostratigraphy of the Ipswichian pollen assemblage biozones to be described.

STRATIGRAPHY AND PALAEOBOTANY OF SECTION B–C

The section B–C (figures 16 and 17) shows a channel, over 100 m wide at the line of section, cut in underlying gravels and containing organic sediments. The section has been described by Markham (1967) and Phillips (1976). The channel is lined by a solifluction diamicton, and the filling above this shows a sequence c. 1.5 m thick with lower and upper beds of sandy mud, separated by thin fluviatile sands and gravels. The organic sediments are related to a fluviatile environment, and contain molluscs, indicating moving water. The estimated height of these organic deposits is 23.3–24.8 m O.D. Below the diamicton at the base of the channel in the underlying sands and gravels is the cast of a large thermal contraction crack, probably dating from the time of the previous cold stage and following incision of the Anglian and Wolstonian sands and gravels. No evidence was seen in the lower part of the section for the separation of Anglian from Wolstonian gravels, though in their coarseness and variability the gravels resembled the Anglian outwash gravels rather than the Wolstonian terrace gravels.

Above the upper mud bed in section B–C is 1.2 m of pale sand with slight signs of stratification, in which Markham (1967) recorded bones as 'fairly common'. This bed is therefore likely to indicate a continuation of the earlier fluviatile conditions of the temperate

Table 5. *Ipswichian pollen assemblage biozones*

	Section		
Substage	B–C (Phillips, 1976)	DDA, DD3	J
III/IV		b. *Pinus*	c. *Pinus–Gramineae* b. *Pinus–Betula–Picea*
IIb	*Pinus–Betula–Alnus–Quercus–Corylus*	a. Gramineae–Compositae–Plantago	a. *Pinus–Betula–Corylus–Quercus–Alnus*
IIa	*Pinus–Betula–Quercus–Ulmus*		

stage. The sand is succeeded by a grey blue clay interstratified with pale sand, disturbed by involutions at the base. This may be Devensian in age. It is at least 0.5 m thick and is overlain by grey stony silt or clay, representing the Devensian solifluction phase on the west slope of the valley. Thus aggradation in the channel reached a level of about 26.5 m O.D.

The palaeobotany of the temperate stage sediments of the B–C sections has been fully described by Phillips (1976). The organic sediments were rich in pollen and plant macroscopic remains. The lower organic bed (section C of Phillips) showed pollen assemblages with high tree pollen frequencies, principally *Betula*, *Pinus* and *Quercus*, ascribed to the Ipswichian regional biozone IIa. The decreasing organic content of the sediments from base to top (table 3) reflects a rise of water table and increasingly fluviatile conditions. The lower mud was overlain by a thin pale sand with black and grey flints, in places cutting into the lower mud. The upper more sandy mud (section B of Phillips) showed high non-tree pollen frequencies, particularly Gramineae, Compositae Liguliflorae and *Plantago lanceolata*, with tree composition (used by Phillips to name the biozone) indicating an age towards the end of regional biozone Ip IIb.

Phillips (1976) discussed the possibility that the high non-tree pollen frequencies of the upper mud bed (section B), with low organic content (table 3), were related to the extension of the sandy floodplain of the river at the time, which increased the area of herbaceous vegetation, and also perhaps to grazing by large mammals. Bones of large grazing mammals were associated with the upper mud bed by Markham (1967), and Phillips (1976) recorded a pollen spectrum assigned to Ip IIb from mud on a *Hippopotamus* bone from the base of the upper mud bed. Further analyses from bones are presented below.

STRATIGRAPHY AND PALAEOBOTANY OF SECTION DD

The DD section (figure 8) gives evidence of a wider extension of the upper mud bed of section B. In this section sandy mud (Bed D) was seen to overlie the gravels of Bed C. The lithology of this mud was similar to that of the upper mud bed of section B, both having scattered angular black flints. The mud was thickest at 3–4 m in the section, thinning to the west as the level of the underlying gravel rose. At 14 m it is represented by a thin horizon of organic sand which continues westwards in the section, where it is much disturbed and associated with sands of Bed F (figures 14 and 15). A similar brown organic stony sand underlies the sandy mud in the eastern part of the section. This horizon is likely to be the remnant of a soil present before a rising water table allowed the deposition of much more organic sediment. In parts of the section, e.g. the DDA section, stratified muddy silt overlies the sandy mud, the topmost part of which has no pollen content. The muddy silt shows an unconformable relation with the sandy mud below and is probably associated with reworking an erosion at the beginning of the deposition of the overlying sands of Bed F (figure 37). The lack of pollen at the top of the sandy mud may be ascribed to weathering. There is much disturbance involving the sediments overlying the sandy mud; this will be discussed later in the description of Bed F.

Two pollen diagrams (figure 18) were constructed from analyses of the mud of the DD section. The first was DD 3, and later, when the similarity with the upper mud in section B emerged and the stratigraphical significance of the similarity was realised, a further section, DDA, was analysed. The DDA diagram is from samples taken on an excavated slope, making the apparent depth in the diagram greater than the vertical depth and giving closely spaced sampling. The two diagrams are similar to an extent, but differ in detail. The differences may be associated with disturbance of the sediments by post-depositional factors (figure 37); these could include the disturbances involving the overlying sediment, and also disturbance by trampling by the vertebrate fauna known from this period.

Though differing in detail, the two diagrams share a simple major zonation as follows:

> b. DD 3 5–35 cm. *Pinus* p.a.b
>
> DDA 10–95 cm. *Pinus* p.a.b
>
> a. DD 3 35–60 cm. Gramineae–Compositae–*Plantago* p.a.b
>
> DDA 95–110 cm. Gramineae–Compositae–*Plantago* p.a.b

Gramineae–Compositae–*Plantago* p.a.b.   This biozone is associated with the more inorganic sediments in the lower part of the sections, as indicated by the loss-on-ignition measurements in table 3. The tree pollen frequencies are low; *Pinus* is most abundant, with very low frequencies of *Ulmus, Quercus, Alnus, Carpinus, Picea* and *Corylus*. The non-tree pollen shows much variety, the most abundant taxa being Gramineae, Cyperaceae, Caryophyllaceae, Compositae Tubuliflorae and Liguliflorae, *Plantago lanceolata* and *Ranunculus*. The assemblage indicates 'weedy' grassland and local wetter habitats. It is of a type associated at other sites with the presence of large mammals (Turner, 1975).

***Pinus* p.a.b.**   This biozone covers a period of increasing *Pinus* pollen frequencies, but the total tree pollen frequency never reaches very high levels. The sediment is much more organic than in the lower part of the section (table 3). There is a reduced frequency of thermophilous trees, but *Picea* maintains a nearly continuous presence. Amongst the non-tree pollen, *Calluna* increases its presence, and Gramineae, Cyperaceae, Caryophyllaceae, Compositae Liguliflorae, Cruciferae, *Ranunculus* and *Succisa* are well represented. Pollen of Cruciferae and *Polygonum bistorta/viviparum* increases towards the top. The assemblage continues to indicate the presence of grassland with wetter habitats available and with indications also of heath in the later part of the biozone. In DDA the change to this biozone is accompanied by a short period of higher frequencies of Cyperaceae pollen, probably indicating flooding of the soils associated with the previous biozone.

A list of macroscopic plant remains from the DD Ipswichian sites is given in table 6. The DD moss sample comes from the upper part of the DDA section. The assemblage generally indicates local open herb vegetation, with many taxa indicating drier (*Cerastium, Saxifraga* cf. *granulata*) or damp grassland (*Ajuga, Cardamine*) or wetter habitats (*Nasturtium, Ranunculus sceleratus, Veronica*). The abundance of *Carex* spp. at the base (DDA 90–105 cm) is associated with the high pollen frequencies at the same level. The macroscopic assemblages give an indication of local plant communities similar to that given by the assemblages described by Phillips (1976) from the upper sandy mud at section B: 'plants of dry sandy ground and wet open habitats'. Phillips' records include macroscopic remains of *Leontodon autumnalis* and *Plantago* cf. *lanceolata*, both taxa of pollen types recorded in the pollen diagrams.

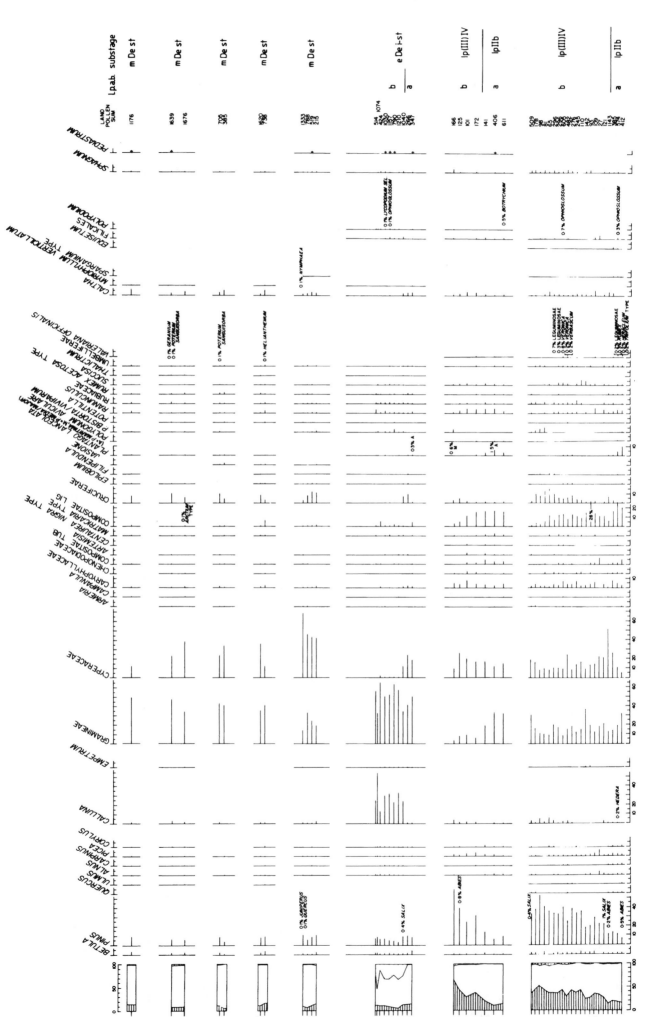

Figure 18. Pollen diagrams of Ipswichian (sections M, CC3, CC4, PP, HH) and Devensian stadial (sections DDA, DD3), Devensian interstadial (section GG) and Devensian stadial (section GG).

Table 6. *Frequency of macroscopic remains from DD sections*

Taxon	Type of remains	DDA 20–35 cm	DDA 50–70 cm	DDA 90–105 cm	DD3 05–35 cm	DDA moss sample
Ajuga reptans	n		1	1		
Caltha palustris	s	1				
Cardamine cf. *pratensis*	s	1				
Cardamine sp.	s	3				
Carex spp.	n(+u)	10	20	60(+29)	7	10
Caryophyllaceae	s		3	3		
Cerastium cf. *arvense*	s	13	6			
C. cf. *fontanum* subsp. *triviale*	s		2	2		
Cruciferae	s			2		
cf. *Eleocharis* sp.	n				1	
Gramineae	car	3	2	8	1	
Hypericum tetrapterum	s			3		
Hypericum cf. *tetrapterum*	s					1
Juncus spp.	s			51	1	4
Lychnis flos-cuculi	s		4	3		
Luzula cf. *multiflora*	s					6
Luzula sp.	s	6	13	22	12	
cf. *Luzula* sp.	s					1
Montia fontana	s				1	
Nasturtium microphyllum	s				1	
Ranunculus acris/repens	a			6		
R. cf. *repens*	a					1
R. cf. *flammula*	a			2		
R. sceleratus	a			1		
R. subg. *Ranunculus*	a		1			2
R. spp.	a			3		
Rumex cf. *acetosa*	f,per	3				
Saxifraga cf. *granulata*	s			40	6	
Saxifraga sp.	s					3
Veronica anagallis-aquatica/catenata	s			1		
Bryophyta	st	*	*	*	1	*
Cristatella mucedo	sta		2	8		2
Carbonised wood	fr			7		

Note:

Abbreviations for type of macro remain and symbols for abundance: a, achene; b, bud, stem with buds; bsc, bract/bud scales; c, capsule; ca, calyx; car, caryopsis; csc, cone or catkin scale; cv, capsule valve; d, drupelet; e, embryo; ep, ephippium; f, fruit; fl, flower; fr, fragment; fst, fruitstone; fv, fruit-valve; l, leaf; lf, leaf fragment; m, megaspore; n, nut, nutlet; o, oospore; p, pod, per, perianth; pet, petiole; s, seed; st, stem fragment; sta, statoblast; u, utricle; +, present; *, many; ∞, very numerous.

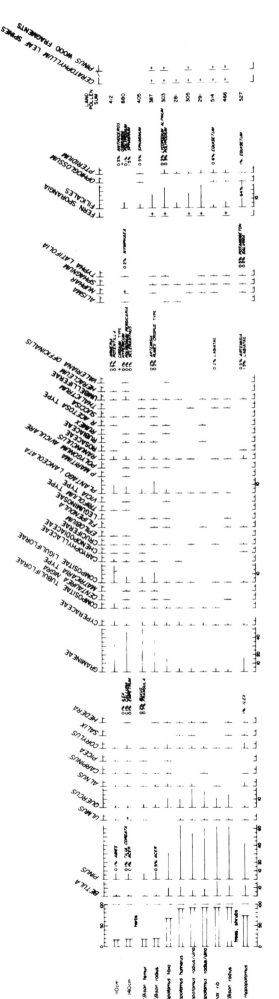

Figure 19. Ipswichian pollen spectra from sediments within or attached to bones of large vertebrates. Percentages expressed as percentage of total pollen excluding aquatics.

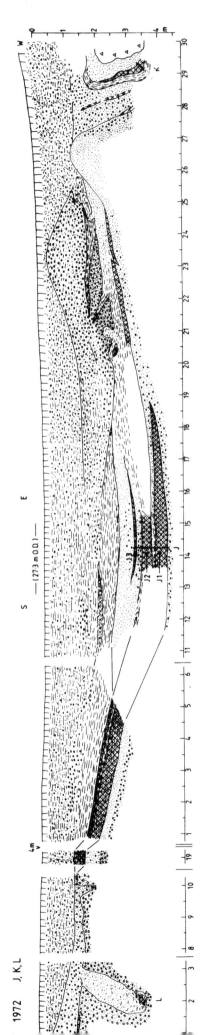

Figure 20. Section J–K–L (north–south–west).

The sequence of the DD sandy mud thus probably represents the accumulation of organic sediment in marshy conditions, with nearby dry sandy ground, an interpretation supported by the nearness of the sandy mud basin to its margin, as shown in the DD section.

The lower biozone of the DD pollen diagrams is similar to Phillips' diagram from the upper mud of section B. This similarity is shown by the top two spectra in the further pollen diagram in figure 19, one from the base of DDA, the other from the upper mud at section B. It will be seen that both the base of the DDA and the upper mud bed show high non-tree pollen frequencies, especially Gramineae, Compositae Liguliflorae and *Plantago lanceolata*, that the tree pollen genera are similar, and both show a similar low organic content (table 3).

The base of the DD sandy mud is therefore correlated with the upper mud of section B, some 200 m to the south-east. The DD diagrams show, however, a slight recovery of regional forest and loss of thermophilous forest genera in their upper biozone, probably indicating climatic deterioration in the later part of the temperate stage.

Phillips (1976) correlated her section B biozone with the regional biozone Ipswichian IIb, so the same age is indicated for the DD biozone a. The age of biozone b is problematic. In the absence of high tree pollen frequencies, correlation with regional pollen zones in the later part of the Ipswichian is difficult. The Ipswichian III biozone with *Carpinus* was recorded by Phillips (1976) at Swanton Morley, but at Roosting Hill the sediments with high non-tree pollen assemblages appear to continue into a time with rising *Pinus* pollen frequencies and the development of heath, which may indicate a time in the late Ipswichian, possibly Ip IV. The assemblages resemble in tree pollen content and non-tree pollen variety those from a late part of the Ipswichian recorded in pollen diagrams at Wing, Leicestershire (Hall, 1980) and at Histon Road, Cambridge (Sparks & West, 1959), both placed in Ipswichian IV.

STRATIGRAPHY AND PALAEOBOTANY OF SECTION J–K–L

The J–K–L section of 1972 (figure 20) is 200 m to the south of section B–C, on a lower part of the slope of the valley and partly at right angles to it. The section shows a depression cut in stony sand, with a width in line of section of about 30 m, and with a base at about 23 m O.D. The filling is unlike that of the B–C channel. At the centre of the basin (at 14 m, J) the filling consists of two beds of muddy silt separated by sand (J 20–100 cm), overlain by a further bed of stony sand, a thin more organic sandy mud (J 0–5 cm), and grey silty clay with a shallow basin of organic sandy silt at its surface. The highest part of the filling preserved is in the southern part of the section at about 25.5 m O.D. The topmost bed is a solifluction diamicton of grey stony silt, the base of which cuts into the upper part of the channel filling.

The sediments of the depression filling are unlike and more inorganic (table 3) than those of the lower mud bed of the section B–C channel filling and the palaeobotany is also very different, the sediments being poor in macrofossils. Particle-size distributions of sediments from the three horizons analysed for pollen (J 1, J 2, J 3; figure 41) show that the two lower horizons are silt rich, while the uppermost (J 3) is more sandy. This uppermost horizon (J 0–5 cm) is also more organic and contains a richer macroscopic flora. It is likely that the depression is closed and the sequence represents sedimentation in shallow water in a pond drying out from time to time, with more fluviatile conditions associated with the J 0–5 cm level.

The stratigraphy at the margins of the depression is more complex and is affected by disturbances of periglacial origin. To the north, the surface of gravel beneath the channel is incised by a large wedge-like structure containing brown sand and silt and an erratic of sandy mud at the base (section L).

To the south, in the east–west part of the section, a diapir of the lower sand has been thrust upwards, resulting at section K in a near-vertical arrangement of the horizon of grey silt and sandy mud seen to the north at 12 m, and in much disturbance of the depression filling (fault at

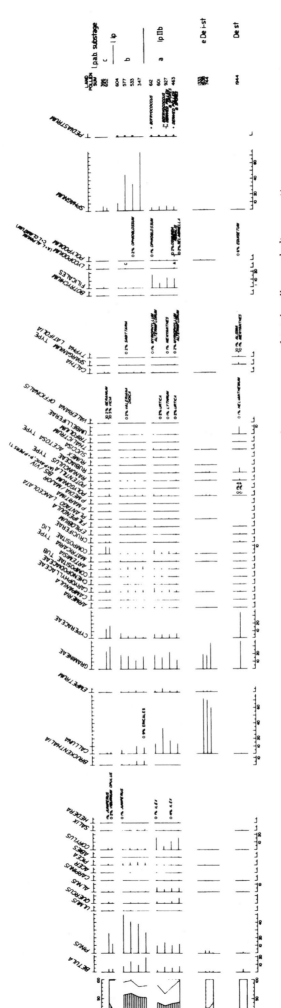

Figure 21. Ipswichian and Devensian pollen diagrams, section J–K–L. Percentages expressed as percentage of total pollen excluding aquatics.

23 m). The injection of the diapir precedes the final solifluction phase which led to the deposition of the diamicton of grey stony silt over the channel, but an earlier period of solifluction involving gravel, resulting in the red-grey clayey gravel below the grey stony silt, appears to be disturbed on the downslope to the east of the diapir. This gravel is at the same height as the gravels of section R–V (figures 24 and 28), and so is probably associated with the same Devensian aggradation. The red gravel seen at the north of the section (0–3 m, section L) has at its base organic streaks, which appear to derive from the depression filling. The wedge-like structure appears older than the gravel, since the gravel plays no part in its filling.

Since the pollen diagrams of J, K and L are all different, it becomes important to place them in a stratigraphic order. The depression filling at J appears to be oldest. The disturbed K sequence of grey silt and sandy clay overlying sands and gravels with thin seams of organic sediment mirrors the sequence in the centre of the depression between 2 m and 2.5 m depth. The K pollen diagram is therefore placed later than the J pollen diagram. The age of the section L pollen spectrum is more uncertain. The sandy mud of L must predate the formation of the wedge-like structure, which formed before the overlying gravel could take part in its filling. This would place the L sample as earlier than the gravel but later than the depression filling.

Of the three sections J, K and L that were studied palaeobotanically, K is related to a later interstadial and L to a later stadial. Results from these two sites will be described in a later section.

The J pollen diagram (figure 21) is through three organic horizons of the depression filling: the two lower muddy silt and the uppermost sandy mud. The three horizons show different pollen assemblages:

a. J 65–95 cm. *Pinus–Betula–Corylus–Quercus–Alnus* p.a.b.   This basal part of the diagram shows about 40% tree pollen, with *Betula, Pinus, Quercus, Alnus* and *Corylus* present in significant quantity. The non-tree pollen is predominantly *Calluna*, Gramineae, Cyperaceae, Caryophyllaceae and Compositae Liguliflorae. If we exclude the possibility of significant parts of this assemblage being reworked, always to be considered for pollen assemblages in highly inorganic sediment (table 3), then this assemblage represents a partly forested regional vegetation, with the tree genera indicating an Ipswichian IIb age. The macroscopic plant remains from this level (table 7) are poor with only *Juncus bufonius*, abundant *Juncus* spp., *Typha* sp. and *Selaginella*. Sedimentation in shallow water with much inorganic sediment influx may explain the nature of these assemblages. Though the tree genera represented and the Gramineae pollen frequencies are similar to those of Ipswichian IIb in the B–C section, there is a difference in the much greater frequency of pollen of *Calluna* in the J section. This is likely to be related to the greater inorganic content of the J sediment containing a greater proportion of inwashed sediment with its soil pollen bank.

b. J 20–50 cm. *Pinus–Betula–Picea* p.a.b.   The pollen assemblages at this level differ from those of the preceding part of the diagram by a reduction of the pollen of thermophilous trees and *Calluna*, and increase of *Pinus* and *Picea* pollen, and the appearance of *Bruckenthalia* pollen. The differences indicate a change in the regional forest vegetation and a later part of the temperate stage, the assemblages bearing a resemblance to the *Pinus* p.a.b. of sections DD 3 and DDA (figure 18). The macroscopic plant remains from this level are again very poor, with cf. *Betula, Juncus* spp., *Typha* and *Chara* (table 7).

c. J 0–5 cm. *Pinus*–Gramineae p.a.b.   The assemblage at this level shows a decrease in pollen of *Pinus* and *Calluna* and an increase in the non-tree pollen, particularly Gramineae, Cyperaceae and Compositae Liguliflorae. The assemblage of macroscopic plant remains (table 7) is richer than in the less organic sediments below and contains cf. *Artemisia sp.* (*Artemisia* pollen present), *Carex* spp., Cruciferae, abundant seeds of *Juncus* spp., cf. *Lemna, Luzula* sp.,

Table 7. *Frequency of macroscopic remains from Ipswichian and Devensian sections*

		Ipswichian							Early Devensian										Middle/Late Devensian					
					stadial				interstadial								stadial		stadial					interstadial
		J			X		S		N	A	UA	T			GG	R	M		CC	XX	RR	AA		H
Taxon	Type of remains	75-90 cm	30-40 cm	0-5 cm	30-60 cm	0-20 cm	60-80 cm	5-20 cm	20-50 cm	45-65 cm	5-20 cm	25-40 cm	15-25 cm	0-14 cm	0-10 cm	5-15 cm	15-25 cm	0-10 cm	0-10 cm	90-130 cm	20-40 cm	70-90 cm	40-60 cm	5-15 cm
Alismataceae	e				40		7																	
cf. Alisma sp.	a,e									1	1													
cf. Armeria sp.	fl fr																							
cf. Artemisia sp.	a			6	1											2								
Betula pendula	fr																							1
Betula sp.	fr						4																	14
Betula sp.	csc						4																	2
cf. Betula sp.	fr		2																				18	
Carex cf. aquatilis	n																						2	
C. cf. nigra	n											10					2							
C. cf. rostrata	n																52							
C. cf. rostrata	n+u																4							
C. spp.	n			42							2	141	1		23	25	20		45	4	1	5		
C. spp.	e											28	4			4			9					
cf. Carpinus betulus	n fr													1										
Cerastium sp.	s											1												
Cruciferae	s			4							1													
cf. Glyceria fluitans	car				132															5				
Gramineae	car										2												5	
cf. Gramineae	car																						1	
Helianthemum cf. canum	s																1							
Juncus bufonius	s	2																						
J. spp. incl. J. bufonius	s				8																			
J. spp. incl. J. cf. effusus/inflexus	s	44	8	159		8	113	131	15	155	143				2	98			3			6		
J. spp.	s			1								3	3											
cf. Lemna sp.	s																							
Luzula cf. multiflora	s																					1	2	
Luzula sp.	s			11			3									13							1	
Lychnis flos-cuculi	s											9												
cf. L. flos-cuculi	s											1												
cf. Lythrum salicaria	s										1													
Minuartia verna	s																					1		
Montia fontana	s				1							3												
Plantago major	s										2													
cf. Picea abies	lf																							4
Potamogeton alpinus	fst																				1			
P. natans	fst																							
P. sp.	fst																			1				
Potentilla anserina	a				7																			
P. sp.	a																							
cf. Potentilla/Fragaria sp.	f			2																		1		
Ranunculus acris/repens	a															4								

	type											
R. flammula	a								1			
R. subg. Batrachium	a	12	7	20		83	3		1	6		
R. subg. Ranunculus	a		1		1					1		
R. sp	a		2			2						
Rumex acetosella group	n											
R. cf. acetosella group	n,f			1					2			
cf. R. sp.	n							1				
Salix sp.	lf				1	4						
	s											
Saxifraga sp.	s			1								8
Stellaria cf. crassifolia	s		3	3			3					
Thalictrum alpinum	a		1									
Typha sp.	s							5		3	2	1
Urtica dioica	a	4					1					
Vaccinium oxycoccus	s											
V. sp	s		1									
Viola sp.	s							19				
cf. Viola sp.	s		1									
Zannichellia palustris	a				1	1	1				7	2
Selaginella selaginoides	m		11	5	12	7	∞			*		3
Bryophyta	st + l		*	*	*	+	+		7		+	
Sphagnum spp.	l / lf						+		*		+	
Characeae	o	+		3	4	+			+			2
Chara	o		1					2			1	
Nitella	o			11				1		2		
Bud scale	bsc									1		
Carbonised wood	fr			1					*	*	+	2
Wood	fr								*			
Cristatella mucedo	sta					1		1		1	
Daphnia	ep					10		5		5	1	2

Note:
For abbreviations of type of remains see table 6.

cf. *Potentilla/Fragaria* sp., *Saxifraga* sp., *Typha* sp., and *Selaginella selaginoides*. The pollen and macro assemblages reflect an increase in the local plant communities of a herbaceous element, with better preservation of locally derived macroscopic remains than in the lower sediments. The age is probably again a later part of the temperate stage.

POLLEN ANALYSES FROM BONES

Further pollen analyses of sediment within or attached to bones are given in the pollen diagram in figure 19. The bones are in the Norwich Castle Museum collection, and were collected in abundance at the time of the original observations on the sections B–C in 1964. The analyses show that the bones are clearly either associated with the lower mud bed and its higher tree pollen percentages (section C) or, as in the case of 3A and 6A, with the upper mud bed (section B), and its high non-tree pollen percentages, including high Gramineae, Compositae Liguliflorae and *Plantago lanceolata*. The top of the pollen diagram shows analyses from the base of DDA and the upper mud bed of Phillips' section B for comparison, and the *Hippopotamus* analysis from Phillips (1976) is shown at the base. The analyses show that *Hippopotamus* is recorded in Phillips' regional biozone Ip IIa, with *Elephas* and *Bos/Bison*. In the upper mud, ascribed by Phillips to Ip IIb, *Bos/Bison* occurs. One *Hippopotamus* bone, 2B, has a pollen spectrum intermediate between these two groups.

It is evident that a rich fauna occupied the Roosting Hill landscape in Ipswichian II, and possibly into III times, as at Swanton Morley (Phillips, 1976). The possible effects of large mammals on vegetation has been considered by Phillips (1974), Turner (1975) and Stuart (1976, 1986). Grazing effects promoting the spread of grassland may account for the characteristic high non-tree pollen assemblages already considered. *Plantago lanceolata* is a plant characteristic of grasslands, often associated with grazing and the drier parts of alluvial meadows and pastures (Sagar & Harper, 1964; Behre, 1981; Groenman-van Wateringe, 1986), where it may grow with plants in the pollen taxon Compositae Liguliflorae (e.g. *Leontodon*, *Taraxacum*). The pollen spectra bear a resemblance to those relating to Iron Age and Medieval grasslands, as described in the Upper Thames valley by Lambrick & Robinson (1988). The analyses in figure 19 establish a relationship between the mammal genera concerned and the vegetational history.

PALAEOGEOGRAPHY

The B–C section, with its fluviatile sediments, cuts across the Ipswichian valley, and the DD section shows marginal organic sediments of the valley. Figure 7 reconstructs the position of the valley, following a course to the north to join the present main valley of the Whitewater valley north of Roosting Hill. Unlike other Ipswichian valley sequences, e.g. that at nearby Swanton Morley in the Wensum valley (Coxon *et al.*, 1980), the Ipswichian sediments are not overlain by fluviatile Devensian gravels (5 m at Swanton Morley), but by sands and silts and then the final solifluction diamicton of the Devensian. This indicates that the Ipswichian channel was abandoned by the main river, which then followed a course east of Roosting Hill, a course which may have already been in existence in the Ipswichian. The stratified fine sediments above the Ipswichian organic sediments in section B–C will then have formed in the channel occupied by quiet water in late or post-Ipswichian times. The channel was evidently abandoned before the time of deposition of the Devensian gravels associated with solifluction diamictons seen in the sections J–K–L and R–V, since these gravels are not seen in the covering deposits of the Ipswichian valley. Possibly the channel diversion is related to solifluction on the valley slopes south-west of Roosting Hill.

CHAPTER FIVE

The Devensian (cold) Stage

The Devensian (or last cold) Stage lasted some 100 000 years, ending 10 000 years ago. It is a complex period, with evidence in southern Britain for severe climates from the nature of the sediments, biota and periglacial phenomena. Within the period there are, however, interstadial episodes of less severe climate when boreal forest tended to replace herb vegetation. Such interstadials are found in the Early Devensian and in the latest part of the Late Devensian, the Devensian late-glacial. In East Anglia the periglacial regime is accompanied in the Late Devensian by an ice advance which reached north Norfolk about 18–20 000 years ago. It is in the Late Devensian that radiocarbon dating can begin to be applied to obtain a better chronology for the identified sedimentary events and biota.

The complexity of the Devensian is reflected in the sequence at Beetley, where a variety of sediments and biota can be related to the evolution of the landscape, aided by the application of radiocarbon dating. The resulting synthesis shows a more detailed Devensian sequence than at any other site in East Anglia.

Many sections exposing Devensian sediments were seen during the working of the gravel pits at Roosting Hill. They extended from the north in the floodplain of the Whitewater southwards along the western side of the valley as far as Whin Covert.

The sections are isolated from one another, and, as will be seen, they are complex and not easy to relate to each other. Several apparently isolated depressions containing organic sediments were recorded, some bordered by structures of periglacial origin. The depressions may have originated through periglacial processes, through subsidence, or they could be associated with aggradation in channels previously eroded, as with the temperate stage sediments of sections B–C and DD.

In addition to the complexity of stratigraphy, there is also complexity in terms of the taphonomy of the pollen and plant macro assemblages. Since the depressions may be isolated, there can be purely local taphonomic factors producing variation in the assemblages, as with the temperate stage sediments already described. The pollen floras are generally substantial and are recorded in *Carex* peats, in sandy organic sediments and in limnic organic silts and clays. All these have very different sedimentary origin, which has a bearing on the interpretation of the pollen assemblages. There is also the possibility of reworking to take into account in such inorganic sediments. The macro floras (table 7) of these sediments are generally poor. They are distinguished by large numbers of *Juncus* seeds and, apart from the assemblages in *Carex* peat, little else. The genus *Juncus* as a whole indicates damp habitats, and, together with the inorganic nature of the sediments concerned (table 3), the frequency of the seeds suggests shallow water or drying pond accumulation with varying degrees of inwash of inorganic sediments and varying water levels. This paucity of macros contrasts strongly with the richness recorded from the temperate stage sediments (sections B–C, DD) and from the later, Devensian, woodland substage (section D–E–F–G figure 22) described by Phillips (1976).

The complexities of the stratigraphy and palaeobotany are a result of the complexity of Devensian processes which occurred on the west side of the Whitewater valley and which

Table 8(a). *Late and Middle Devensian pollen assemblage biozones*

	Section	Radiocarbon age BP	Pollen assemblage biozone
Late Devensian interstadial	H	c. 11.9 ka	Betula
Middle/Late Devensian stadial	BB 1, 2	c. 16.5 ka	Cyperaceae–Gramineae
	WW–XX	c. 27.2 ka	Cyperaceae–Gramineae
	RR	c. 31.7 ka	Cyperaceae–Gramineae
	AA		Cyperaceae–Gramineae
	M	c. 38.0 ka	Cyperaceae–Gramineae
	CC, HH, PP		Cyperaceae–Gramineae
	R		Cyperaceae–Gramineae

Table 8(b). *Early Devensian pollen assemblage biozones*

	Section D (Phillips, 1976)	Section A	Section S, U	Section N, P	Section T	Section GG	Sections Z, K	Section V	Section X
stadial	d. Gramineae	e. Gramineae–Calluna–*Pinus–Betula*							
	c. *Pinus–Betula–Picea*–Ericales–*Sphagnum*	d. *Pinus–Betula–Picea*–Calluna			c. *Calluna*	b. Gramineae–*Calluna*	*Calluna*–Gramineae	Cyperaceae–*Calluna*	
		c. *Pinus–Betula–Picea*	*Pinus–Betula*–Gramineae	b. *Pinus–Betula*–Gramineae	b. *Pinus*–Cyperaceae	a. Gramineae–Cyperaceae			
interstadial	b. *Pinus–Betula–Picea*	b. Cyperaceae–*Pinus–Betula*		a. *Pinus–Betula*	a. *Calluna*–Gramineae–Cyperaceae				
		a. *Pinus–Betula–Picea*							
stadial	a. Gramineae								Gramineae

originate in climatic change, the course of aggradation and down-cutting in the valley, and the processes of solifluction, solution and hydrology that are associated with the variation in lithology of the underlying Anglian sediments and the presence of the Chalk.

The positions of the sections, both on the western slope of the valley and in the floodplain, are shown in figures 2 and 7. The pollen diagrams from the Devensian sites indicate both times of woodland (interstadial) and times of herb vegetation, described here as stadial rather than full-glacial, as seen in the summary of the succession of Devensian pollen assemblage biozones in table 8. Table 10 summarises Devensian sedimentary events.

Sites with woodland vegetation will be described first, starting with the most complete vegetational sequence of section D–E–F–G, then the shorter sequence of section A and the sequences at sections S–T–U–Z–V, K, GG, and N–P, JJ and Q. There is no evidence in these sections for more than one forested interstadial. This interstadial has pollen assemblages similar to those of the interstadial at Chelford, Cheshire (Simpson & West, 1958); a correlation with that interstadial was suggested by Phillips (1976) in her discussion of the palaeobotany of the section.

The sections showing herb vegetation will then be described; these include sections M, CC, HH, NN, PP north-west of Roosting Hill, sections L, W, X and WW–X in the area south-west of Roosting Hill, and the floodplain sections to the north, sections BB, AA, RR and H.

A number of radiocarbon determinations have been obtained from Middle and Late Devensian and Flandrian organic sediments. The measurements were made by Dr V.R. Switsur's laboratory in the Subdepartment of Quaternary Research, University of Cambridge. They are described in detail in Appendix IV. In the following text the determinations are not given in full each time they are considered, but are abbreviated to a *circa* age BP without probabilities.

SECTIONS D–E–F–G AND A

SECTION D–E–F–G

This section is shown in figure 22. It was recorded and sampled in 1969 and was later studied by Phillips (1976). The section shows a shallow basin filled by a series of organic sediments, 1.6 m thick in the centre, overlying a pale sand. Observations by J. Webb (pers. comm.) at the time showed that this basin with its organic sediments lay within a depression in the Anglian gravels, a depression filled with about a metre of 'interbedded light green silts and pure white sands' which graded up at the margin of the depression into stony white sand. The basin with organic sediments thus rests within this filling. Webb also observed slumping of the organic sediments towards the centre of the basin, which is likely to be associated with its possible origin through subsidence. Although it is impossible to relate the section to the sections that show temperate stage deposits at sections B–C and J, it is probable that the interbedded silts and pale sands at the base of the depression are equivalent in age to the similar sediments that were seen at about the same level to overlie the temperate stage deposits at B–C. This would confirm that the organic deposits of the D–E–F–G section are later than the temperate stage, as originally suggested by Phillips (1976).

The filling of the basin at D is as follows:

0–100 cm	light brown stiff sandy mud
100–125 cm	irregular paler muddy silt, top 10 cm level with gravel at E
125–135 cm	light brown coarser stiff mud

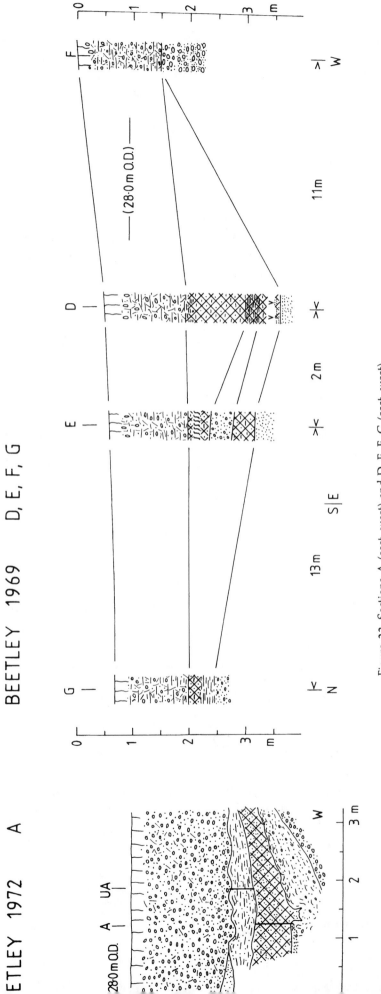

Figure 22. Sections A (east–west) and D–E–F–G (east–west).

135–150 cm   brown wood peat with *Betula* wood and *Pinus* needles

150–160 cm   darker brown-black mud with laminations to white sand

below 160 cm white sand

At E the filling is:

0–7 cm   sandy mud

7–20 cm   moss–sedge peat

20–35 cm   sandy mud

35–70 cm   pale stony sand and gravel

below 70 cm mud

At G, at the eastern margin, 30 cm of mud overlay greenish silt. At F, upslope beyond the western margin, 1.5 m of grey stony silt directly overlay the Anglian gravels. This grey stony silt, a solifluction diamicton, forms a sheet over the basin, as seen in the section.

The basal wood peat of D must have formed near water level. At an earlier time limnic mud was deposited, indicating a hydroseral change as the pool filled. Later, water level rose, and finer organic sediments were formed, accompanied by the inwash of silt at D 100–125 cm and sand and gravel at E 35–70 cm. The derivation is from marginal stony sand observed in the section by Webb (pers. comm.). The moss–sedge peat at E is a pool margin sediment. These changes of water level may be associated with climatic change or hydrological changes associated with subsidence.

The organic sediments in the basin were rich in plant remains, both pollen and macroscopic plant remains. These were fully described by Phillips (1976), who identified four pollen assemblage biozones as follows:

Section D: d. 0–90 cm   Gramineae p.a.b.

           c. 90–130 cm   *Pinus–Betula–Picea*-Ericales–*Sphagnum* p.a.b.

           b. 130–150 cm   *Pinus–Betula–Picea* p.a.b.

           a. 150–160 cm   Gramineae p.a.b.

Section E: 15 cm   Cyperaceae–Gramineae p.a.b.

The richness of the assemblages in these sediments enabled Phillips to make a detailed reconstruction of the local and regional vegetational history. An early grassland phase was followed by a forest phase with *Pinus*, *Betula* and *Picea*, all three genera represented by abundant macroscopic remains. Reduction in forest cover followed, with an expansion of herbs and ericaceous shrubs, indicating the spread of grassland and heath. Especially notable in this phase is the abundance of seeds of *Bruckenthalia spiculifolia*. Soon after this, solifluction into the pool occurred, with the deposition of stony sand and silt. In the final phase forest and heath disappeared, with a change to open herbaceous vegetation.

Philips (1976) discussed the age of this sequence, possible alternatives being a late part of the Ipswichian temperate stage or an early Devensian interstadial. One problem was whether the Gramineae p.a.b. at the base was related to local or regional vegetation. If local, then the high non-tree pollen ratio would not indicate regional grassland, and there would be no evidence for the beginning of an interstadial. Phillips showed that the forest pollen assemblage was similar to that at Chelford (Simpson & West, 1958), and she favoured a correlation with an early Devensian interstadial.

From the stratigraphical evidence discussed above, the organic sediments of the section

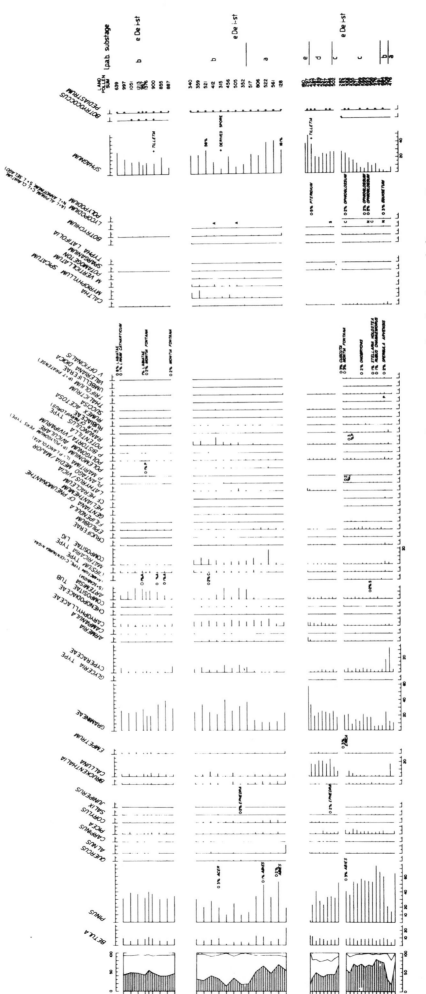

Figure 23. Devensian interstadial pollen diagrams, sections A, P, N. Percentages expressed as percentage of total pollen excluding aquatics.

D–E–F–G certainly appear to be later than the Ipswichian temperate sediments of sections B–C and J. In addition, the palynological evidence from section A, to be discussed in the next section, indicates that the grassland phase at the base of D is regional, because the basal sediments of A, unlike those of D, show the development of a high Cyperaceae phase likely to be related to local vegetation developing around the pool near the beginning of its existence. It would be expected that a similar Cyperaceae peak would be present at D if the pollen assemblage was of local origin. The conclusion is that a whole interstadial is shown by the vegetational history at D.

SECTION A

This section, drawn in figure 22, shows a further depression in the gravels to the south of the D–E–F–G section. The depression is steep-sided and is lined by light brown sandy clay with small flints, probably a solifluction diamicton. Above this is a thin horizon of pale sand, contorted together with the upper part of the sandy clay. The filling is composed of two units. The lower is a brown sandy mud, 65 cm thick at the centre of the depression, very sandy in the bottom few centimetres, then much more organic from 40–50 cm. The central part shows contortion with the underlying pale sand and sandy clay. The upper unit has a sharp junction with the unit below and is a grey brown sandy clay. The depression filling is covered first by a pale and then by unstratified red brown sandy silty gravel, 1.8 m thick. The junction between this gravel and the underlying sandy clay is contorted, suggesting that the gravel is of solifluction origin from upslope Anglian gravels.

The depression at A may also be the result of subsidence. The basal sequence is similar to that in the section D–E–F–G, with a basal few centimetres of very sandy mud, the much more organic mud, then sandy mud. Changes in water level, following an initial rise, are again indicated. The change from the lower to the upper unit is a result of increased inwash of inorganic sediment into the water body in the depression. The loss-on-ignition measurements in table 3 show these changes.

The pollen diagram from section A (figure 23) is divided into two parts, A and UA, corresponding respectively to the lower and upper units described above. The following pollen assemblage biozones are distinguished:

UA	e. 0–7 cm	Gramineae–*Calluna*–*Pinus*–*Betula* p.a.b.
	d. 7–37 cm	*Pinus*–*Betula*–*Picea*–*Calluna* p.a.b.
	c. 37–45 cm	*Pinus*–*Betula*–*Picea* p.a.b.
A	c. 0–52 cm	*Pinus*–*Betula*–*Picea* p.a.b.
	b. 52–62 cm	Cyperaceae–*Pinus*–*Betula* p.a.b
	a. 62–65 cm	*Pinus*–*Betula*–*Picea* p.a.b.

The basal sample of A reflects the presence of forest with *Pinus*, *Betula* and *Picea*, but the assemblage soon changes to one with much increased Cyperaceae pollen, occurring in the more organic sediment near the base of the filling. The change reflects a greater representation of local pollen from around the pool in the depression. *Caltha* is also at its highest frequencies in this biozone. Higher in the A section the assemblage returns to a forest one dominated by *Pinus*, *Betula* and *Picea*. *Bruckenthalia* is also present, with Gramineae, low frequencies of *Calluna* and Cyperaceae. There is a wide range of other non-tree pollen taxa, including Caryophyllaceae, *Artemisia*, Compositae, *Filipendula*, *Polygonum bistorta*/ *viviparum* type. The whole assemblage indicates forest with opportunities for more open ground herbaceous communities and heath.

With the change to the more inorganic sediments in the UA diagram, the *Calluna* frequencies greatly increase, the Gramineae frequencies increase slightly and the *Pinus* frequencies decrease. This change is best explained by greater inwash increasing the influx of *Calluna* pollen, rather than by gross vegetational changes. In the topmost biozone the tree pollen frequency drops to below 25%, and Gramineae percentages increase strongly, with a rise of *Armeria* and *Plantago* pollen frequencies. These changes indicate a change towards herbaceous vegetation and disappearance of forest.

Two samples of the lower sandy mud were analysed for macroscopic remains, a 5–20 cm and 45–65 cm (table 7). In the lower sample, although numerous plant fragments and pieces of carbonised wood were present, the only identifiable remains were very numerous *Juncus* spp. seeds and a cf. *Alisma* embryo. The upper sample was also poor, with the same abundance of *Juncus* seeds, but had a few examples of other taxa, including cf. *Alisma*, *Carex* spp., cf. *Picea abies* and *Sphagnum*. These records, poor though they are, confirm the general interpretation based on the pollen assemblages and sediment, indicating deposition in a shallow pool.

It is possible to relate the UA and A pollen diagrams to Phillips' diagram from the D section, as shown in table 8. The *Pinus–Betula–Picea* p.a.b. of A corresponds to the similar biozone of D, Phillips' p.a.b. b. The similarity extends to the non-tree pollen, e.g. *Filipendula*, *Gentiana pneumonanthe*, *Succisa*. Phillips' p.a.b. c, which shows a decrease of *Pinus* pollen and an increase of heath and herb pollen, corresponds to the *Pinus–Betula–Picea–Calluna* p.a.b. of the upper sandy clay of section A. The proportion of *Calluna* in A, however, is much greater; this difference is probably related to inwash from a soil pollen bank. The topmost biozone of A, showing greatly increased frequencies of Gramineae pollen, may then correspond to the Gramineae p.a.b. d of Phillips, marking a time of climatic deterioration and expansion of herbaceous communities. The great differences in the plant macro content of D and A must relate to the differences in the sedimentary environment, with the paucity (except for *Juncus*) in A a result of shallow water sedimentation, perhaps in an ephemeral pool. It should be added that there is no stratigraphically based reason to suppose that this correlation between section A and F–D–E–G is not correct.

SECTIONS R–Z, X, V, W

STRATIGRAPHY

These sections are shown in figure 24, 25, 26, 27 and 28. The outline stratigraphy is simple, but the details are complex. The underlying gravel, containing wisps of pale chalky till, is Anglian. Overlying this gravel is a series of organic sediments and fluviatile sands, the whole covered by a sequence of solifluction diamictons with intervening beds of gravel.

Section X–V–W (figure 25 and 28) is the simplest of the two sections. Section V shows a depression in the Anglian gravel containing 6 cm of sandy mud overlain by pale stratified sands. The preserved part of the depression was seen to be a semicircle, presumably a closed depression on the surface of the gravel. Northwards the organic horizon is traceable at a higher level as a darkened surface of the underlying gravel. This appears to be the remnant of a soil, covered by about 3 m of grey stony silt, separated in two beds by an intervening red ill-sorted flint gravel. The sequence in the V depression indicates deposition of organic sediment near water level, with a later rise in water level resulting in the deposition of fluviatile sands. Upslope from V is the section W, which showed the edge of a depression containing 35 cm of brown sandy mud, with interbedded stony sand resting on brown sandy silt and covered by 40 cm of grey-brown stony sand and clay, a solifluction diamicton. This depression again rests on Anglian gravels.

ETLEY 1972   R, S, T, U, X, Z

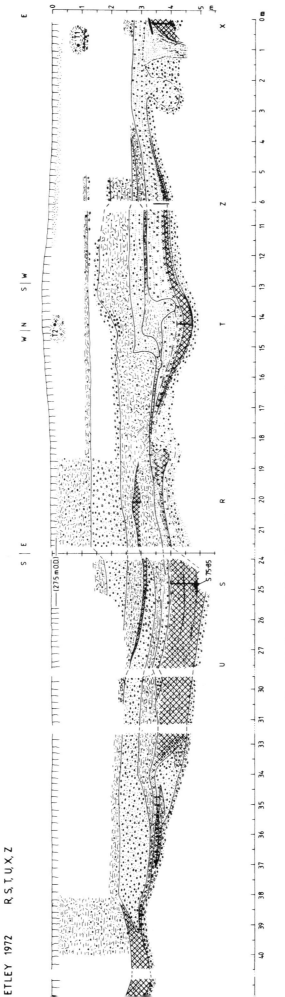

Figure 24. Section R–X (north–south, east–west, north–south, east–west).

Figure 25. Section Z, X (figure 24), 0–6 cm. Anglian gravels at base (A), Devensian interstadial organic sediments (E), Devensian solifluction diamictons (G) and gravels (Ga). Cast of thermal contraction crack to right of spade. At right, Devensian stadial organic sediments.

Figure 26. Section R, T (figure 24), 15–21 m. Anglian gravels at base (A), Devensian interstadial organic sediments disturbed by a diapir (E), Devensian solifluction diamicton (G) and gravels (Ga).

Figure 27. Section T (figure 24), 15 m. Devensian fluviatile sands overlying interstadial organic sediments (E).

To the north end of the X–V–W section the soil horizon extends at section X (figure 25) over a 60-cm thick bed of black sandy mud which rests on Anglian gravels and dips steeply north before being cut off by a wedge-like structure with vertical lineation of pale sand and silt. The sandy mud of X is evidently older than the organic filling of the depression at V, and between the two disturbance and warping of the older organic sediments took place through injection or thermal contraction cracking.

Turning now to the contiguous section R–S–T–U–X–Z (figure 24), the southern end of this section (0 m) overlaps with the north end of section X–V–W. A balloon-like structure containing muddy sand and gravel lies under the soil level and may be associated with the disturbances which cut off the mud of section X, already discussed. A further depression in the underlying gravels is seen from 4 to 17 m on the section. This is filled by a thicker sequence of organic sediments than the depression V. At section T (figures 26 and 27) the sequence is as follows:

0–14 cm	dark brown stratified mud and sand
14–26 cm	brown sandy mud
26–38 cm	sedge peat
38–42 cm	very sandy mud with small flints

At section Z, nearer to the margin of the depression, the filling is sandy mud at the base, then interstratified mud and pale brown sand. The depression filling is covered by fluviatile sediments, stratified muddy sand, which can be seen to extend over the soil to the south. The mud is likely to be reworked from the older organic sediments. The depression filling appears to belong to the same surface as the filling at V, with the lower level at T resulting in a thicker organic sequence with sedge peat.

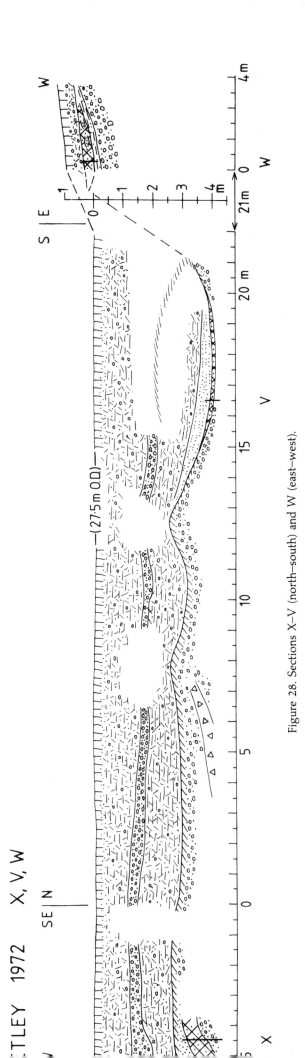

Figure 28. Sections X–V (north–south) and W (east–west).

At the north-east end of the T depression a diapir (figure 26) disturbs and interrupts the marginal sediments of the depression at 18–19 m. The surface of the underlying gravel here is sandy and develops an upper purple and lower pale colour, possibly relating to pedogenic processes at the time. To the north-east of the diapir, this purple horizon dips down, and a larger depression (sections S, U) is seen filled with nearly a metre of stratified brown silty mud, with thin sand seams in the marginal part and thin clay seams in the central part. The final sediment of the depression filling is a light brown sandy clay, and there then follows sand and gravel relating to the fluviatile sediments overlying the depressions at T and V.

The north margin of the depression at 34 m is steep-sided with the adjacent silty mud containing stones derived from the margin. Here there are again disturbances of the underlying sand and gravel. To the north (34–39 m) the shallow marginal part of the depression is marked by grey-brown organic sand, similar to that on the southern margin at 20–21 m.

The stratification of the depression filling, and the presence of seams of sand and clay in the marginal and central parts respectively, indicates the filling of a water body with intermittent introduction of sand and clay. A similar feature, with a thin bed of sand in Flandrian limnic organic sediments was seen in the depression filling at site Q to the south in Whin Covert, where a seepage or spring occurs. For comparison, loss-on-ignition measurements and particle-size distributions for both the S–U depression and Early Flandrian limnic sediments at site Q are given in table 3 and figure 41 respectively. It is unlikely that the filling of the large depression at S–U involved stream flow of any magnitude, and it is better regarded as sediment formed in a quiet water body subject to the introduction of inorganic sediments, perhaps by changes in spring flow or overbank flow from the Whitewater stream of the time.

The sediments overlying the sequences described above include interdigitating sands, gravels, and solifluction diamictons with an associated organic sediment. The details are described in a later section.

PALAEOBOTANY

The pollen diagrams from the sections V, Z, T, U and S described above are shown in figure 29.

Section T (figure 27)   The pollen diagram from section T shows clear variation of the pollen spectra related to variations of sediment. The following pollen assemblage biozones are distinguished:

    c. 0–26 cm   *Calluna* p.a.b.

    b. 26–35 cm   *Pinus*–Cyperaceae p.a.b.

    a. 35–40 cm   *Calluna*–Gramineae–Cyperaceae p.a.b.

The basal spectrum at 40 cm is in very sandy mud with small flints, and shows an assemblage with low tree pollen, and with substantial frequencies of *Calluna* and Gramineae and a variety of herbaceous pollen taxa. With the deposition of sedge peat (26–38 cm) the *Pinus* and Cyperaceae values rise, and *Calluna* is very low at 30 cm, where there is a *Pinus* maximum of 34%. Low inwash of inorganic sediment into the sedge peat, results in low frequencies of *Calluna*. With the return to sandy mud and interstratified sandy mud and sand, *Calluna* frequencies rise to high values (80%), suppressing the *Pinus* frequencies. Loss-on-ignition measurements through the T sequence are shown in table 3. The pollen diagram indicates the changing importance of inwash, as indicated by the loss-on-ignition results, to the total pollen assemblage, and provides a key for the interpretation of other similar pollen diagrams.

Macroscopic plant remains were analysed from three levels of section T (table 7). The basal sample was from the sedge peat and contained a considerable flora. The most abundant

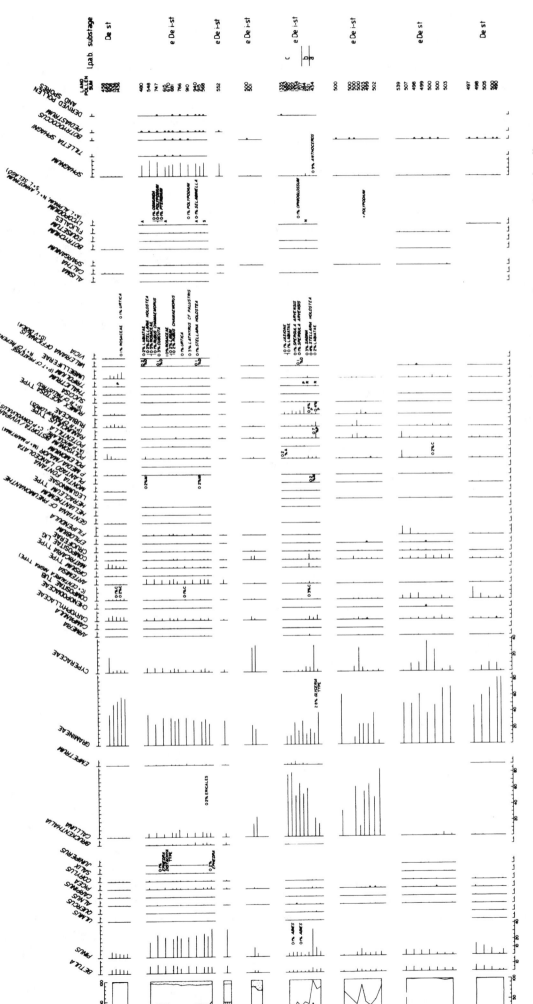

Figure 29. Devensian stadial (sections W, X, R) and interstadial (sections Z, T, V, S, U) pollen diagrams. Percentages expressed as percentage of total pollen excluding aquatics.

remains were those of *Carex* spp., including *C.* cf. *rostrata*; also present were *Cerastium* spp., *Luzula multiflora*, *Lychnis flos-cuculi*, *Montia fontana* and *Ranunculus* sections *Batrachium* and *Ranunculus*. No seeds of *Juncus* were found. The assemblage is one compatible with sedge communities growing under telmatic conditions, as suggested by the sediment and stratigraphy. At the level of 15–25 cm, the flora was much poorer, containing few *Carex* spp. remains, Gramineae caryopses and *Juncus* seeds. At this level the sedge peat is replaced by sandy mud as water level rises, and more inwash of inorganic sediment is associated with higher *Calluna* pollen frequencies, as described above. At the topmost level, 0–10 cm, of stratified sand and mud, only half a carbonised nut of *Caprinus* and wood fragments, some carbonised, were found. The sediments here show the beginning of fluviatile sedimentation, with much coarse reworked organic material.

Section Z   This diagram is from a shallower and marginal part of the T depression. It shows a sequence of pollen spectra with low tree pollen (*Pinus, Betula*) and high frequencies of *Calluna* and Gramineae pollen. There is a peak of Cyperaceae pollen at 27 cm. The sediments are largely interstratified mud and pale sand, and the high *Calluna* frequencies are again related to inorganic inwash indicated by the sediment.

Section V   The sandy mud formed in the V depression again shows a pollen assemblage with low tree pollen (*Pinus, Betula*), and high frequencies of *Calluna*, Gramineae and Cyperaceae. The high frequencies of the last probably relate to local vegetation around the shallow pool. The assemblage bears comparison with that from section GG 1 (figure 30), described below, where high *Calluna* frequencies are again related to inwash and shallow pool conditions.

The three diagrams from T, Z and V thus measure pollen sedimentation in autochthonous and allochthonous sediments of shallow pools on the irregular surface of older gravels.

Sections S, U   The pollen diagram from section S covers the main part of the limnic filling of the much deeper depression at S–U. The additional section U samples were taken to examine the assemblage near the contact with the underlying gravel, well below water level at section S. There is no great change of the pollen assemblage through the S–U diagrams. It can be described as a *Pinus–Betula*–Gramineae p.a.b. Tree pollen percentages are around 40%; *Pinus, Betula* and *Picea* are the main trees represented, with traces of other taxa, some thermophilous. Shrubs include *Ephedra, Bruckenthalia* and *Calluna*, the last with frequencies not so high as in the sediments that show more inorganic inwash at T, Z and V. The non-tree pollen is mainly Gramineae, but there are consistent low frequencies throughout of Cyperaceae, Caryophyllaceae, *Artemisia*, Compositae, *Filipendula*, and *Polygonum bistorta/viviparum*, as well as low frequencies of many other herb taxa, including *Armeria* and *Rubus chamaemorus*. Aquatic or marsh plants are represented by *Alisma, Caltha* and *Sparganium*. There are considerable frequencies of *Sphagnum* spores throughout the diagram.

This pollen assemblage is very similar to the *Pinus–Betula–Picea* p.a.b of section A, except that the percentages of Gramineae are higher. It is correlated with that biozone (table 8).

Two samples from section S were analysed for macroscopic plant remains (table 7). The lower sample, 45–60 cm, contained little except abundant *Juncus* seeds. The upper sample, besides abundant *Juncus* seeds, contained *Betula* fruits and cone scales, remains of *Luzula* sp., *Potentilla* sp., *Rumex* cf. *acetosella* and *Typha* sp. This paucity of macros, with the abundance of *Juncus*, seems typical of silty limnic sediments of the Devensian, as it is of such sediments in the Ipswichian (section J), and may be characteristic of assemblages surviving in silt-rich sediments with fluctuating water levels (see Appendix II). Particle-size distributions from section J and the interstadial sections S and N are shown in figure 41. The N analysis resembles in the high silt

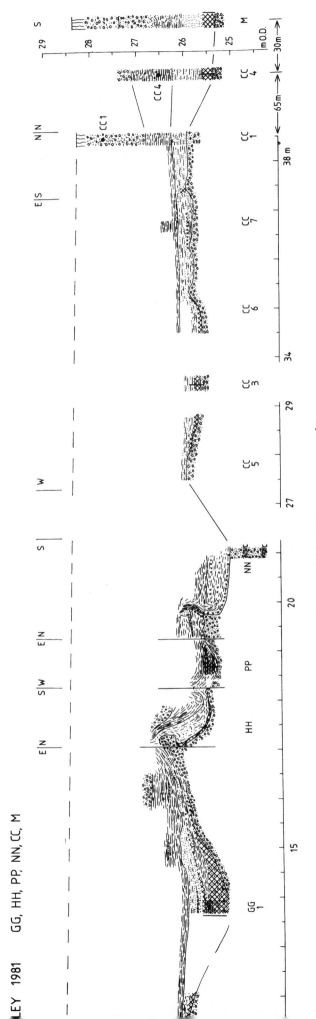

Figure 30. Sections GG to M (west–east–south).

content those from the J 1 and J 2 analyses, while the basal S analysis (S75–85 cm) resembles with its increased sand content that from the uppermost sample J 3.

The uniformity of the S–U pollen assemblage contrasts with the varied assemblages of sections A, D, and T, particularly because there are apparently no changes at the base of the sequence, as are seen in these other sections, where they have been related to climatic or hydroseral changes. The uniformity is likely to be related to rather rapid deposition. The absence of change in the whole diagram indicates stability of vegetation and environment; the absence of change at the base indicates sudden formation of a sedimentary basin in which conditions remained the same. Such a basin could be formed by subsidence, known to occur in the area, or alternatively in a meander cut-off of a gravel floodplain subject to overbank silting. Since the underlying gravels are of Anglian type and contain till, the latter explanation is unlikely.

We can then envisage the relationship of the R, S, T, U, X, Z and X, V sections as follows. Subsidence in the area during the Devensian interstadial already recognised formed a shallow pool. At a late stage of the filling of this pool, which took place rather rapidly, a rising water level allowed deposition of organic sediment in neighbouring shallow depressions (T, V). Finally, stratified sands were deposited over the organic sediments in the depressions, indicating a period of more fluviatile conditions, probably associated with the floodplain of the time.

SECTIONS K AND GG

These two sections are similar to sections V, Z and T described above. Section K (figure 20) at the south-western end of the long J–K–L section, shows a near vertical side of a diapir with grey silt overlain by sandy mud, as described previously. A similar *in situ* sequence is seen in the centre of the J depression at 12–13 m in the section. No macroscopic plant remains were present in the organic sediment, but pollen was present and the pollen diagram (figure 21) shows a *Calluna*–Gramineae p.a.b. which is similar to the assemblages with *Calluna* from sections V, Z and T, representing a part of the interstadial already identified in the pollen diagrams discussed above. The high *Calluna* frequencies are again associated with low organic content. The stratigraphical relation to the temperate stage sediments of section J confirm that the interstadial represented is later than the temperate stage.

Section GG is part of the long section DD to the north-west of Roosting Hill shown in figures 8 and 30. The section shows a small depression in gravels which are filled first with 8 cm of coarse sandy mud, followed by 40 cm of stratified silts and silty mud. The pollen diagram from the section at GG 1 shows pollen spectra dominated by non-tree pollen, especially Gramineae and *Calluna*. Two pollen assemblage biozones are present:

    b. 8–40 cm   Gramineae–*Calluna* p.a.b.

    a. 0–8 cm   Gramineae–Cyperaceae p.a.b.

At the base, high Cyperaceae values accompany the most organic part of the filling. As the inorganic content increases (table 3), so do the values of *Calluna* pollen. A variety of other non-tree pollen types is present, but only Cruciferae and *Caltha*, near the base, are in any quantity. The sequence represents the filling of a small depression, first by coarse mud, with the pollen spectra reflecting local conditions, and the late increase of *Calluna* is associated with inwash of inorganic material, probably from local soil destruction. As with the other pollen diagrams with high *Calluna* pollen frequencies, *Betula* and *Pinus* both have a low frequency. The

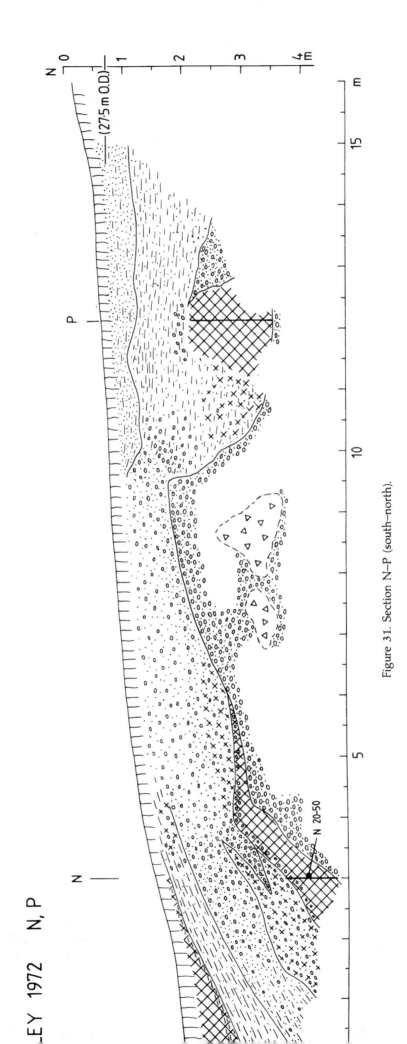

EY 1972   N, P

Figure 31. Section N–P (south–north).

assemblages are related to the period of the interstadial in which heath developed. The list of macroscopic plant remains from GG 0–10 cm (table 7) is short. Nuts of *Carex* spp. are present, with *Juncus* spp. seeds, *Zannichellia* achenes and a *Sphagnum* leaf. The finds add little to the palaeoecology, apart from confirming the origin of the basal organic mud in a shallow pool with *Carex* present.

SECTIONS N–P, JJ, AND Q

Section N–P is a section (figure 31) excavated on the north-western margin of the large depression VV. The depression contains Flandrian organic sediments and is the source of a spring. Section JJ (figure 32) is a section excavated on the south-west margin of the same depression. Section Q (figure 33) is an excavation on the south-west margin of a further but smaller depression to the south, also filled with Flandrian organic sediments and the source of an intermittent spring.

SECTION N–P

This section (figure 31) shows a very irregular sequence, with beds dipping steeply into the present large depression at the southern end, and a rise of Anglian gravel with large till clasts in the centre. The outline sequence is the basal gravel with till, followed by brown sandy or silty mud in two separate places, followed by gravelly sand, grey clayey silt and finally pale silty sand at the top of the northern end of the section. It is clear that at the southern end of the section there has been subsidence leading to the dip of the sequence seen today. The brown silty mud (the particle-size distribution is shown in figure 41) at N is overlain by stony mud, associated with reworking as subsidence took place. Gravelly sand overlies this, then grey silt and clay, followed by Flandrian organic telmatic sediments. The section P shows the same general sequence, with sandy gravel then blue-grey clayey silt overlying organic sediment. But this section has been greatly disturbed, possibly by what may appear to be diapirs of gravel, containing till clasts, a few metres to the south and to the north.

The sands and gravels and clayey silt overlying the organic sediment at N and P may be equivalent to the sands and gravels and stony silt overlying the organic sediments to the north at section S–U. They are at similar levels. Such a correlation would imply that at the N–P section a basin or basins of organic sediment, of the same age as the organic sediment at section S–U, were covered by sands and gravels either related to fluviatile deposition or solifluction, and later by stoneless clayey silt, an equivalent of the solifluction stony silts over S–U. Subsidence followed solifluction, but the central diapir and the associated disturbances appear to overlap the time of solifluction, since gravel merges into the clayey silt at 10 m in the section, but the main movement evidently took place before deposition of the clay.

Pollen diagrams (figure 23) have been obtained from sections N and P. The P diagram covers 1.2 m of brown sandy mud, but, in view of the associated disturbances, it is uncertain whether this covers 1.2 m of original vertical depth of sediment. Two pollen assemblages are present:

    b. 0–75 cm   *Pinus–Betula*–Gramineae p.a.b.

    a. 75–120 cm   *Pinus–Betula* p.a.b.

The lower p.a.b. has over 50% tree pollen, while the upper has reduced *Pinus* pollen percentages and higher percentages of pollen of Gramineae, Cyperaceae and especially *Artemisia*. The section N pollen diagram shows a *Pinus–Betula*–Gramineae p.a.b., similar to the upper biozone of section P, but with higher frequencies of *Pinus* and lower frequencies of

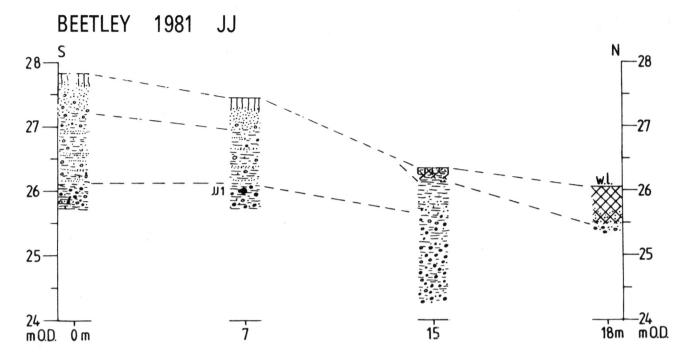

Figure 32. Section JJ (south–north).

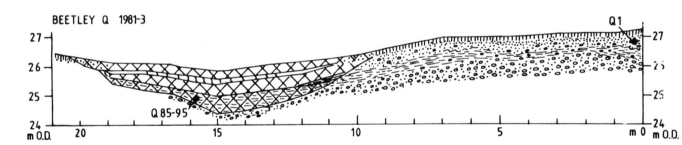

Figure 33. Section Q (north–south).

Caryophyllaceae, Compositae Liguliflorae and *Ranunculus* type. The N and P diagrams generally resemble the S–U diagrams, and may be placed in the latter half of the interstadial. The correlation is shown in table 8, suggesting also correlation with Phillips' p.a.b. c, showing a rise of Gramineae and *Artemisia* in this her *Pinus–Betula–Picea*–Ericales–*Sphagnum* p.a.b. The N and P diagrams also resemble the S–U diagram in the lack of change at the base, suggesting a similar origin of their basins to that of the S–U depression. Particle-size distributions of the limnic sediments of N and S are compared in figure 41.

A macroscopic remains analysis from section N, 20–50 cm, gave again a very poor assemblage, with only *Juncus* spp. and *Typha* spp. (table 7).

SECTION JJ

This section (figure 32) was excavated on the south-west margin of the VV depression, and was found to be completely different from the N–P section. At the base was a blue-grey silty clay with flints and chalk pebbles, followed by up to 1.2 m of brown sandy silt with sand lenses. Where the slope towards the depression started, 0.5 m of grey-brown clayey silt occurred at the top of the section under the soil. This is likely to be equivalent to the grey silt and clay near the top of sections N and P. There was 60 cm of brown cover sand with small flints at the top of the southern part of the section. The blue-grey clay with chalk pebbles at the base has a particle-size distribution similar to the solifluction diamictons of section R–V, as is seen in figure 41, and may be interpreted as a soliflucted Anglian till. Upslope, brown and grey chalky till was found in boreholes at a higher level O.D. (figure 38). The brown sandy silt, in places stratified with sand, appears to be a slope wash or solifluction diamicton with a similar derivation. The till may have provided a source for the grey clays and silts of section N–P and for the solifluction diamictons in the sections further to the north.

SECTION Q

This section (figure 33) was excavated south-west from the centre of the closed depression Q, to the south of the JJ trench. Grey and brown gravel was seen to be overlain by grey and mottled clay and silt, with 0.5 m brown cover sand with small flints at the surface (the particle-size distribution is shown in figure 41). To the north the beds slope down towards the depression, and there is grey gravel with chalk pebbles at the base of the trench. A depth of 1.5 m of Flandrian organic sediments fill the depression. The gravel in this section is at about the same height as the Devensian gravels in section R–V to the north and the Bed C Wolstonian gravels further north. The grey and mottled clay and silt are probably equivalent to the similar sediment in sections JJ and N–P, occurring beneath cover sand.

The pollen diagram from the Q depression (to be published elsewhere with a more detailed consideration of Flandrian vegetational history) shows at its base a high non-tree pollen assemblage followed by a *Betula* p.a.b. These biozones are Devensian late-glacial and Flandrian I in age. They indicate that deposition started in the hollow in Devensian late-glacial times, suggesting that the hollow was formed or that water could accumulate in it from Devensian late-glacial times. It is likely that the hollow originated through subsidence in Late Devensian times.

SECTIONS WITH HERB POLLEN ASSEMBLAGES NORTH-WEST OF ROOSTING HILL

STRATIGRAPHY

The sections to be described are shown in figures as follows:

Section M (figure 34)

Sections DD, GG (figure 8)

Sections GG, HH, NN, PP, CC (figure 30)

Sections CC, EE, TT (figure 9)

These sections show an extensive basin of deposition (Bed F) (see figure 7) sealed by a solifluction diamicton (Bed G). There is the following sequence of sediments, resting on black flint gravels at CC and on coarser Anglian gravels at M. Approximate O.D. heights are given.

Grey silty clay, with yellow sand at margins of basin	26.5–27.0 m O.D.
Yellow and brown sand and silty sand, with thin fining-up sequences at base	25.6–26.5 m O.D.
Black coarse detritus mud	25.4–25.7 m O.D.

The sequence represents the filling of a shallow depression or valley as water level rose. First, coarse detritus muds (high loss-on-ignition; table 3) were deposited in shallow water in the lows of the irregular surface of black flint gravels (sections M, CC 3, CC 4, HH), with a more silty and stratified facies at PP. A further rise brought a more inorganic sediment into the depression, with changes in flow giving stratified sands and silts containing thin fining-up sequences, and small faults (figure 35). The variable flow indicated may be associated with variable spring flow into a valley leading to the Whitewater. At a later stage of the filling more lacustrine conditions obtained, with the deposition of silty clay (particle-size distribution at CC 4, DD 6 in figure 41), except at the margins of the depression (EE, figure 9, 68 m; DD, figure 8, 15–36 m), where sands were deposited. Particle-size analyses of these (DD 1 to 4; figure 41) indicate they are cover sands, related to aeolian activity at the edge of the depression. The aerial photograph (figure 36) shows the extent of the fine lake sediments in this area, indicated by the darker shade of damper ground. The surface of the filling reaches about 27 m O.D., near the height of the low terrace gravels (Bed C) to the north (figure 13). The section M (figure 34) shows a gravelly sand overlying the silty clay, with an irregular base and disturbed by festooning; this horizon appears to be a solifluction diamicton originating from the nearby rise in the underlying Anglian gravels.

The depression appears to be cut into the terrace gravels at its northern end (EE, figure 9). In the section DD the filling sediments overlie Ipswichian sandy muds (figure 37) and also the small younger depression which is filled with organic sediments at GG (figure 30), described above. Solifluction over the whole filling, together with cryoturbation, has produced an irregular surface of the filling (DD, figure 8), with flames of sediment rising into the solifluction diamicton and bent downslope. In addition to these irregularities, deposition of the basal part of the filling accompanies disruption of the underlying sandy muds in DD, resulting in the reworking of the mud into the basal sediments of the filling (figure 37).

A further type of disruption is seen east of the GG section in figure 30. The pollen assemblages of HH, PP, CC 3, CC 4, and M are similar. All these sediments belong to the first

BEETLEY 1972 M

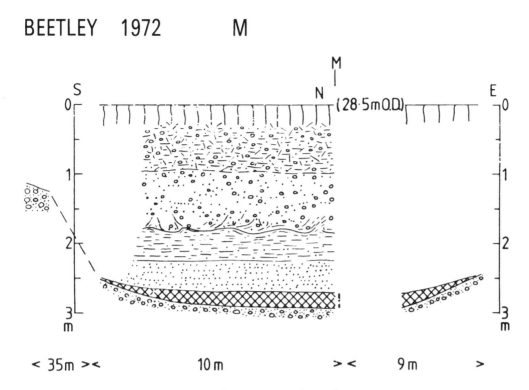

Figure 34. Section M (south–north–east).

Figure 35. Section CC (figure 30), 37 m. Middle Devensian stadial organic sediments (F) overlying Wolstonian gravels and overlain by stratified sands and silts in thin fining-upwards sequences. Devensian lake sediments at top.

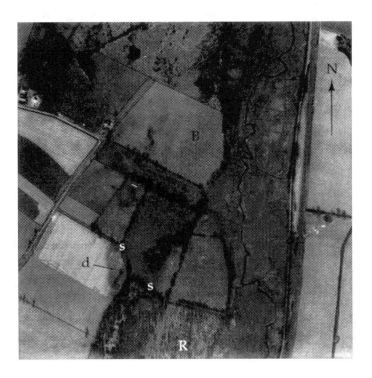

Figure 36. Aerial photograph of central part of study area taken in March 1946, showing extent of damper ground (d) associated with springs (s) and underlain by Devensian lake sediments. To the west is better drained land underlain by Anglian sand and gravel. B, Beetley Terrace; R, Roosting Hill. Crown copyright/RAF photograph.

Figure 37. Section DD (figure 8), 0–3 m. Surface of Ipswichian organic sediments (D) overlain by Devensian lake sediments (F), the upper part of which is more clayey. The interface is much disturbed.

part of the depression filling. GG shows the totally different pollen spectra described above. Between GG and CC 5 (figure 30) a diapir of the underlying gravel has thrust up at 17 m in the section, with a fault and collapse between 19 and 21 m. These disturbances involve the base of the solifluction diamicton, and so took place after solifluction had started. The absence of a basal organic sediment at site NN suggest that collapse played a part in deformation. In recent time a spring rose at about this point, and it appears probable that hydrostatic pressures, associated with subsidence and released near the time solifluction started, are responsible for the dislocations seen.

The origin of the depression remains to be discussed. The relationship with the bed C cold stage gravels and Bed D temperate stage sandy muds indicates an erosional origin after a fall in the base level of the valley. There is no direct evidence for subsidence playing a part. A radiocarbon determination of the coarse detritus mud of section M gave an age of *c.* 38.0 ka BP.

PALAEOBOTANY

Pollen spectra were obtained from sections M, CC 3, CC 4, HH, and PP. They are shown in figure 18. Macroscopic plant remains were abstracted from organic sediments of sections M and CC 4.

The pollen diagrams from the organic sediments at the base of the depression at the five sites are similar, and can be described as a Gramineae–Cyperaceae p.a.b. There are low tree pollen sums, with only *Betula* and *Pinus* present in more than a trace. Non-tree pollen constitutes 80% or more of the total pollen, and is largely composed of Gramineae and Cyperaceae pollen. There is a variety of non-tree pollen taxa present, but only Cruciferae and *Caltha* are present in quantity. Pollen of Caryophyllaceae, Compositae Tubuliflorae and Liguliflorae, *Ranunculus*, Rubiaceae, *Rumex acetosa* type and Umbelliferae is consistently present at all the sites. The pollen assemblages indicate open herbaceous vegetation with much grassland, bordering a shallow pool with sedges and *Caltha*.

Analyses of macroscopic plant remains in the coarse detritus mud were made from two levels at M and from CC 4 (table 7). The lower level at M included abundant remains of cf. *Carex rostrata* and *Carex* spp., corresponding to the high Cyperaceae pollen frequencies at M; *Helianthemum* cf. *canum*, *Montia fontana*, *Ranunculus-Batrachium*, *Sellaria* cf. *crassifolia*, *Vaccinium oxycoccus* and *Zannichellia* are also recorded, representing members of local herbaceous damp ground and aquatic communities. The CC 4 assemblage is much poorer; only *Carex* spp. remains are abundant, with *Juncus* spp. and *Chara*. The macroscopic remains thus confirm the presence of local sedge communities at the time water level started rising in the depression.

SECTIONS WITH HERB POLLEN ASSEMBLAGES SOUTH-WEST OF ROOSTING HILL

The sections to be described are shown in figures as follows:

Section L (figure 20)
Sections X and R (figure 24)
Section W (figure 28)
Section WW (figure 38)

The relative ages of the organic sediments of these sections are shown in table 8. X appears to be the oldest, older than the interstadial of the S and T sections. The age of L is more uncertain; it is older than the post-interstadial gravel of section J–K–L. R is post-interstadial and related to the solifluction diamicton and gravels covering the interstadial sediments of the S and T sections. WW is immediately beneath the final solifluction diamicton which thins out over the floodplain gravels. The age of W is uncertain. It is upslope and isolated from the V section.

SECTION L

The single pollen spectrum from the mud erratic at L in the wedge-like structure in the J–K–L section (figure 20) is shown in figure 21. The pollen assemblage has a very low tree pollen content (< 5%), with the non-tree pollen mostly Gramineae, Cyperaceae, Rubiaceae and Umbelliferae. Other taxa are only present in traces. The assemblage resembles those from section M described above and, like the M sediment, has a high organic content (table 3). It is characteristic of the grassland and herb assemblages of so-called full-glacial type. It is possible that it represents the vegetation associated with the formation of the wedge-like form.

SECTION X

The pollen diagram from 60 cm of sandy mud at section X in the long section R–Z (figure 24) is shown in figure 20. The pollen spectra form a Gramineae–Cyperaceae p.a.b. They show very high non-tree pollen frequencies, with only *Pinus* and *Betula* well-represented amongst the trees. The non-tree pollen is mostly Gramineae and Cyperaceae, the former the most abundant. Pollen of Caryophyllaceae and Compositae is present in low frequency, and pollen of *Filipendula*, *Polygonum bistorta/viviparum* and *Ranunculus* type rise to the top of the diagram. The pollen assemblage indicates surrounding grassland, with changes at the top suggesting shallowing water with local damp ground communities better represented. Macroscopic remains in two samples (0–20 cm, 30–60 cm; table 7) were very scarce, with *Juncus* spp. abundant in both, and cf. *Glyceria fluitans* and Alismataceae embryos common in the lower sample.

The pollen assemblage records a time of herb plant communities before the interstadial and thus in the Early Devensian.

SECTION R

In the long section R–Z (figure 24), a brown sandy mud 20 cm thick and of low organic content (table 3) overlay a solifluction diamicton and gravel at 19–27 m in the section. A pollen diagram from this mud is shown in figure 29. One pollen assemblage is present, a Gramineae p.a.b. The assemblage has less than 15% tree pollen with *Betula* and *Pinus* about equally represented. Most of the non-tree pollen is Gramineae, but also consistently present are Cyperaceae, with higher frequencies at the top, Caryophyllaceae, Compositae–*Matricaria* type, *Polygonum bistorta/viviparum* and Umbelliferae. The assemblage indicates regional grass-land, with a variety of herbs and local damp ground communities. Macroscopic plant remains from the site included cf. *Armeria* sp., *Carex* spp., abundant seeds of *Juncus* spp., *Luzula* sp., *Ranunculus acris/repens*, cf. *Rumex* sp., and *Viola* sp. The pollen and macro assemblages are of interest in being closely associated with a solifluction sheet and therefore likely to demonstrate the regional and local vegetation at a time of post-interstadial solifluction, with plant remains deposited in a shallow pool.

SECTION W

This again is a small depression (figure 28) containing sandy mud overlain by a solifluction diamicton. It is not possible to relate it to the sequence downslope considered above. The pollen assemblage, a Gramineae p.a.b., shows very low but rising tree pollen frequencies, with *Betula* and *Pinus*. The non-tree pollen is largely Gramineae, with Cyperaceae and Compositae Tubuliflorae; less diversity of non-tree pollen is present than seen in the R assemblage described above.

SECTION WW–XX, TRANSECT 13

This east–west section (figure 38) shows a depression or channel filled with silty mud lying in the incised surface of gravels which are associated with the floodplain of the Whitewater at the east end of the section. The stratigraphy indicates a younger age than the interstadial sediments of section R–V, present at section Z (0 m) on the section. A radiocarbon determination of *c.* 27.2 ka BP from the silty mud at XX on the section indicates a Middle Devensian age, so giving this minimum age for the gravel sheet underlying and to the east. Between the silty mud and the overlying solifluction diamicton on the slope of the valley are stratified silts and sands to a height of about 26 m O.D., partly overlapping and partly cut into the silty mud of the depression. The sequence of silty mud and stratified silt and sand indicates a period of ponding and higher water levels in a shallow valley cut in the lower gravels and in the depression containing the silty mud, a valley likely to be related to the nearby spring at site VV, with the ponding possibly related to icing in the valley, as discussed later.

The pollen diagram from section WW (figure 39) shows a Gramineae–Cyperaceae p.a.b. Gramineae percentages are much higher at the base, where the sediment is more silty. As the sediment becomes more organic, Cyperaceae values increase and so do those of pollen of *Ranunculus* cf. *aquatilis*. The sequence represents the filling of a shallow pool. Frequencies of tree pollen are low. *Pinus* and *Betula* are both present, and there are also low frequencies of *Quercus*, not consistent with the assemblage and likely to be of secondary origin. Of the non-tree pollen taxa, there are remarkably high frequencies of *Armeria*, A, B 1 and B 2 types being recorded. Compositae Liguliflorae, Caryophyllaceae, *Plantago media*, *Polygonum convolvulus* are also well represented. The assemblage represents a herb-rich grassland. The macro remains from the nearby section XX (table 7) are poor; they include *Salix*, *Potentilla anserina* and Gramineae, and in the marsh or aquatic category *Carex* sp., *Ranunculus* subg. *Batrachium*, *R. flammula*, *Zannichellia palustris* and Characeae.

FLOODPLAIN SECTIONS

STRATIGRAPHY

This group of sections, H, AA, RR, BB, shown in figure 13, is within gravels aggrading to near the present floodplain level of the Whitewater. These gravels, with possibly older gravels, rest at a depth of 5–10 m on an irregular surface of Anglian chalky till. H, AA and RR are depressions filled with brown sandy detritus mud (loss-on-ignition figures are in table 3) and are interpreted as cut-off meander channels filled with organic sediment. In contrast, the BB sections, BB 1 and BB 2, are parts of the mainly inorganic filling of a very shallow pool on the gravel surface. BB 1 is a facies with laminated drift mud and silt, rich in *Salix* leaves. BB 2 is a more limnic silty sediment with shells.

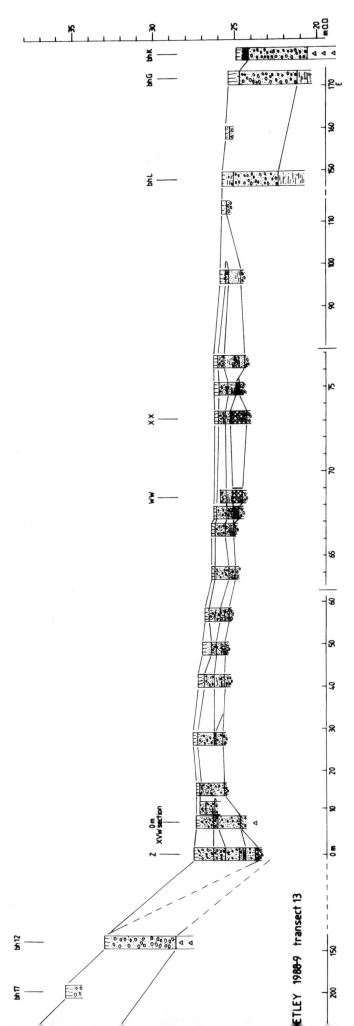

Figure 38. Section (west–east) on transect 13 (boreholes and sections Z, WW, XX).

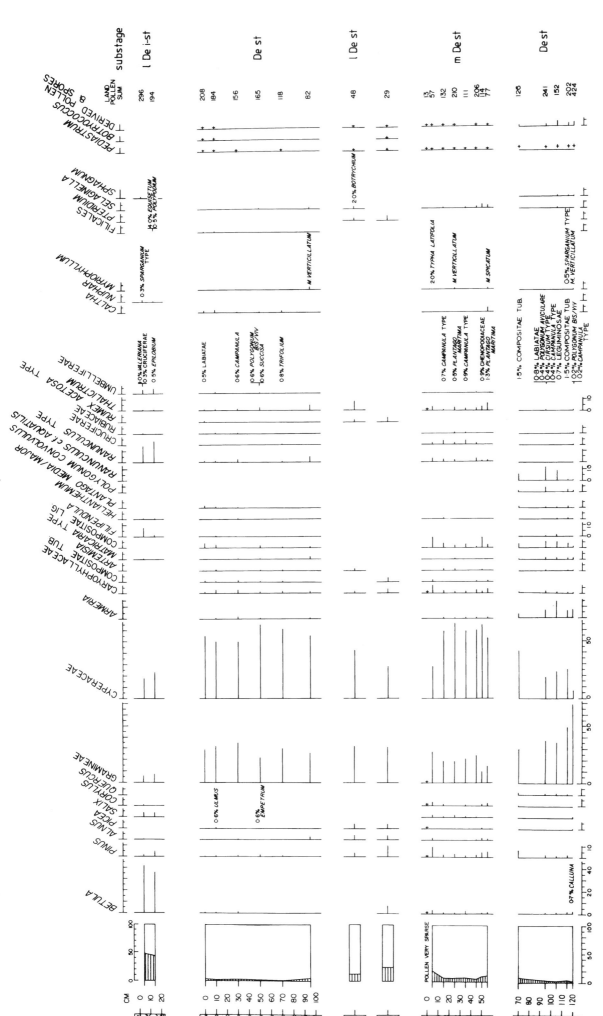

Figure 39. Devensian stadial (sections WW, RR, BB 1, BB 2, AA) and interstadial (section H) pollen diagrams. Percentages expressed as percentage of total pollen excluding aquatics.

Section AA   This showed a wide channel filled at its centre with 98 cm of brown sandy mud with a moss-rich horizon at 85 cm. The channel deposits were overlain by 1.1 m of stratified gravels slightly festooned at the top.

Section RR   This section showed a channel 4 m wide. The centre of the channel contained 7 cm of grey silty clay overlying 50 cm of brown sandy mud, with a pale sandy disturbed horizon at 46 cm, probably the result of sand movement over the floodplain surface at the time. The mud contained scattered small flints, more frequent near the steeper western margin. The channel was overlain by 1.8 m of stratified flint gravel. A radiocarbon determination of mud at depth 16–40 cm gave a result of *c*. 31.7 ka BP.

The uniform filling of the AA and RR depressions indicates that there was no regular flooding of the ponds by mainstream overbank flow during the time of their filling by organic sediment, suggesting limited changes of water level at that time.

Section BB   Slabs of laminated drift mud and silt to 30 cm thick were taken from the section at BB 1, a shallow pond filling overlying coarse sand and gravel and overlain by about 2.4 m of stratified gravel which was chalky at the base. This facies is characteristic of periodic flooding of lows associated with a floodplain. Laterally a deeper water silty facies with shells, BB 2, was recorded. A radiocarbon determination of extracted *Salix* leaves from BB 1 gave a result of *c*. 16.5 ka BP.

Section H   This section showed 20 cm of coarse detritus mud overlying coarse flint gravel in a wide and very shallow channel. The channel was overlain by 1.2 m of ill-sorted sharp flint gravels, taken to be a product of solifluction from the nearby slope of Roosting Hill (figure 43). Alluvium to a depth of 60 cm overlay this gravel. A radiocarbon determination from the channel filling, 5–15 cm, gave a result of *c*. 11.9 ka BP. This result indicates that the organic horizon belongs to the Windermere Interstadial of the Devensian late-glacial (Coope and Pennington 1977), with the overlying gravel a product of the later climatic deterioration of Loch Lomond Stadial times.

If the radiocarbon dates are a true indication of age, then the gravels under the present floodplain started accumulation before *c* 31.7 ka BP. Also the relative depth of the organic horizons is not an indication of relative age. The depths are a guide to floodplain level at the time, so that variation depends on the predominance of aggradation or erosion.

PALAEOBOTANY

Pollen diagrams (figure 39) have been prepared from the organic sediments of all four floodplain sections.

The section AA pollen diagram shows an assemblage with very low tree pollen representation, with the non-tree pollen largely Cyperaceae and Gramineae (Cyperaceae–Gramineae p.a.b.). There is a restricted variety of other herb taxa, of which we may note *Armeria*, Caryophyllaceae, Compositae Liguliflorae, *Helianthemum*, *Plantago*, *Ranunculus* type, and *Thalictrum*. The assemblage indicates the presence of herbaceous vegetation on the floodplain and in the region, with grassland and sedge.

Macroscopic remains were analysed from two levels, 70–90 cm and 40–60 cm (table 7). The assemblages were interesting but not rich. The samples recorded the presence of *Carex* cf. *aquatilis*, *Carex* spp., Gramineae, *Juncus* spp., *Luzula* sp., *Lychnis flos-cuculi*, *Minuartia verna*, *Potentilla anserina*, *Ranunculus-Batrachium*, *Rumex acetosella* group, *Saxifraga* sp., *Stellaria* cf.

crassifolia, *Thalictrum alpinum*, *Vaccinium* sp., cf. *Viola* sp. and *Selaginella selaginoides*. The assemblage adds identifications for the local flora surrounding the cut-off channel, giving a number of species associated with open ground, grassland, damp ground and aquatic communities.

The section RR pollen diagram again shows a Cyperaceae–Gramineae p.a.b. The tree pollen frequency is slightly higher, with *Pinus* most abundant, and low frequencies of *Betula*, *Alnus*, *Salix* and *Corylus*. The presence of the last, which must be reworked, gives an indication of the presence of secondary pollen in the assemblage. The non-tree pollen representation is similar to that in the AA diagram, but with more frequent Caryophyllaceae, Compositae Liguliflorae, Cruciferae and *Thalictrum*. A macro analysis from 16–40 cm showed a poorer assemblage than AA, with *Carex* sp., *Potamogeton natans*, *Ranunculus-Batrachium*, *Salix* sp., and *Selaginella selaginoides*. The vegetation indicated by the RR results is similar to that indicated at the AA section.

The results from the BB sections show a different aspect of the Devensian vegetation. The pollen analyses are much more restricted in variety than those of the AA and RR sections, and the macroscopic analysis from BB 1 is extremely rich. These differences are related to the taphonomy of the assemblages. The BB sediments formed in a shallow pool on the floodplain, and received macroscopic remains from the streams of the time, whereas the AA and RR organic sediments formed in isolation from the streams. The poor pollen assemblages of the BB sediments are related to the same difference, with more reworked pollen and less good preservation of a variety of taxa in the inorganic sediments. Thus the pollen analyses from BB 1 and 2 show a higher frequency of tree pollen (*Betula*, *Pinus*, *Alnus*, *Picea*) than AA and RR, less Cyperaceae pollen and much reduced variety of herb taxa. The assemblages can still be considered to belong to a Cyperaceae–Gramineae p.a.b.

The list of macroscopic remains from the drift mud of BB 1 is shown in table 9. It is the richest Devensian macro assemblage found in the area, a richness associated with its derivation from the catchment of the Whitewater stream in Devensian times. The list in table 9 has been divided into lifeform or habitat groups. Trees are only represented by cf. *Alnus* sp. and *Betula* sp. The most abundant fossil is *Salix herbacea*, though no *Salix* pollen was recorded. *S. herbacea* is a basiphilous species of a wide range of habitats in the arctic, including late snow patches, open slopes and screes and solifluction slopes (see Godwin, 1975). There is a long list of herb taxa, most light-demanding , and most associated with grassland of a basic type (e.g. *Minuartia verna*, *Saxifraga hypnoides/rosacea*) or with damper types of grassland (e.g. *Primula* subg. *Aleuritia*, *Selaginella*). A few taxa are more commonly associated with acidic soils (e.g. *Rumex acetosella* group). The variety must be associated with floodplain gravel habitats and the slopes of the valley, where Anglian chalky tills and outwash gravels and their solifluction products now outcrop. There are also several marsh or fen taxa, but rather few aquatic taxa.

The assemblage is characteristic of the British full-glacial flora, considered by Godwin (1975), with Arctic species represented by *Salix herbacea*, and northern species by *Thalictrum alpinum*. But most of the taxa have a much more wide-ranging distribution at the present day, and are mostly species which are light-demanding.

A mollusc assemblage from BB 2 is described by R.C. Preece in Appendix III.

The section H pollen diagram shows a *Betula* p.a.b, with a radiocarbon determination indicating a Windermere Interstadial age in the Devensian late-glacial. Pollen of *Pinus* and *Salix* is also present, giving a tree pollen total of about 45%. Pollen of Cyperaceae, Gramineae, *Filipendula*, *Ranunculus* type and Umbelliferae forms the non-tree pollen total, probably associated with local wet habitats on the floodplain. In the macro analysis (table 7), there are records of *Betula pendula*, *Potamogeton alpinus*, *Ranunculus* subg. *Batrachium*, *Urtica dioica* and Characeae, again indicating an element of the pool and local vegetation.

Table 9. *Late Devensian macroscopic remains from section BB 1*

Taxon	Type of remains	Frequency	Taxon	Type of remains	Frequency
Trees			*S.* cf. *hypnoides*	s	11
cf. *Alnus* sp.	f	1	*Selaginella selaginoides*	m	140
Betula sp.	f	3	*Stellaria* cf. *crassifolia*	s	4
Shrubs			*S.* cf. *gramineae*	s	1
Salix herbacea	l	127	*Thalictrum alpinum*	a	7
	lf	45	*Helophytes*		
S. sp.	s	1	*Eleocharis palustris*	n	2
	lf	∞	*E.* cf. *palustris*	n	1
	bsc	126	*Eupatorium cannabinum*	a	1
	c	44	cf. *Parnassia palustris*	s	1
	c + st	2	*Ranunculus sceleratus*	a	7
cf. *S.* sp.	b	21	*R. flammula*	a	4
Vaccinium cf. *oxycoccus*	s	1	*Stellaria palustris*	s	1
Herbs of open habitats			*S.* cf. *palustris*	s	1
cf. *Arabis* type	s	3	*Thalictrum flavum*	a	1
Armeria sp.	ca	6	*Hydrophytes*		
Cerastium cf. *arvense*	s	24	*Callitriche* cf. *stagnalis*	d	11
Cochlearia cf. *officinalis*	s	2	Characeae	o	4
Draba incana	s	8 in fv	*Hippuris vulgaris*	a	2
	fv	20	*Myriophyllum spicatum*	n	10
D. cf. *incana*	s	126	*Potamogeton crispus*	fst	3
D. sp.	s	27	*P. filiformis*	fst	2
cf. *Geranium* sp.	s	1	*P. pectinatus*	fst	2
Helianthemum cf. *canum*	l	7	*P. pusillus*	fst	2
H. sp.	s	6	*P.* spp.	fst	3
cf. *H.* sp.	s	1	*Ranunculus* subg. *Batrachium*	a	41
Luzula cf. *multiflora*	s	8	*Zannichellia palustris*	a	6
Minuartia verna	s	3	*Unclassified*		
Potentilla anserina	a	4	*Carex* spp.	n	135
P cf. *tabernaemontani*	a	3	Caryophyllaceae	s	1
P. spp.	a	2	Cruciferae	s	3
Primula subg. *Aleuritia*	s	6	cf. Cruciferae	s	4
P. veris/elatior	s	2	Gramineae	car	23
Rhodiola rosea	s	2	cf. Leguminosae	p	1
Rumex acetosella group	f	10	cf. *Rubus* sp.	fst	1
	f + per	3	cf. *Stellaria* sp.	s	1
Saxifraga hypnoides/rosacea	l	22 rosettes	Bryophyta	l, st	*
	l	66		st + c	4
	pet	25	Pteridophyta	m	3
	st + l	40	*Other*		
			Cristatella mucedo	sta	1
			Daphnia sp.	ep	31

Note:

For abbreviations of type of remains see table 6.

POST-INTERSTADIAL DEPOSITS AT SECTIONS R–V, J–K–L AND N–P

The complex of post-interstadial deposits seen in sections R–V, J–K–L and N–P now requires further description and interpretation.

SECTION R–V

This section (figures 24, 25, 26, 27 and 28), generally parallel to the valley side, is complex, showing alternate solifluction diamictons and fluviatile sediments. Above the interstadial deposits are stratified sands. At Z (4–13 m) a grey stony silt overlies the fluviatile sands. This is the northward extension of the lower grey stony silt of the X–V–W section. It thins to the north, becoming sandy silt at 14 m, where it is involved in an involution containing grey organic sand. This horizon represents movement from the east and south of sediment by solifluction into the area of the depressions T and S–U; the transition to stoneless sandy silt towards the west indicates deposition in water. This sediment is covered by a bed of pale sand and gravel with stratification, indicating renewal of fluviatile conditions. This bed thins to the south at 4 m and to the west at 15 m where it is involved in the involution affecting the older sediments in the central part of the T depression. A further grey stony silt covers these upper fluviatile sediments; to the south this is continuous with the lower grey stony silt of section X–V–W.

To the north, at section R (20 m), this grey stony silt is divided by the thin lens of brown sandy mud, with a transition to the stony silt below and a sharp junction with the stony silt above. This organic horizon represents the sediment of a shallow hollow on the surface of a solifluction lobe. At its northern end the lens rests on red ill-sorted sandy gravel, representing the margin of fluviatile aggradation. At the north end of the section (38–44 m) a further organic horizon overlies the gravel and both are overlain by grey stony silt to the surface. A later red gravel occurs between the grey stony silts along the section X–V–W, extending and thickening into the R–X section, where it is thickest (1.2 m) at 21 m. This upper gravel is succeeded by the upper stony silt, a further solifluction diamicton, common to both sections.

The post-interstadial sediments of the section R–V are thus seen to represent deposition near the margin of a floodplain, with alternating fluviatile sedimentation and solifluction diamictons. The fluviatile sediments become coarser with time, and the solifluction is first from the east and south. Thus aggradation accompanies the onset of solifluction. The section WW and transect 13 (figure 38) are downslope and at right angles to section R–V. They show that the fluviatile sediments and earlier solifluction diamictons (A; table 10) occupy a depression, while the upper diamicton (B; table 10) blankets the present slope. Thus this upper diamicton is later than the period of downcutting. But like the capping diamicton seen north-west of Roosting Hill, it overlies, near the foot of the slope, stratified sediments formed during a period of ponding before the final solifluction phase.

From the section (figure 6) showing the relation of sections on the west slope of the valley, it will be seen that the interstadial sediments of section R–V lie at a low level, allowing the deposition of the earlier solifluction diamictons interdigitating with fluviatile sediments, not seen in the sections to the north-west of Roosting Hill. The origin of the lower level could result from deposition in an area of subsidence, or from a difference of levels associated with confined hollows on the one hand (e.g. sections A, D–E–F–G) and contemporary valley bottom sediments on the other hand.

The relative age of the disturbances seen in section R–V can now be summarised. The wedge-like structure at 1 m (figure 24) appears to post-date the early Gramineae–Cyperaceae

Table 10. *Devensian and Early Flandrian events*

Stage	R–Z area	Floodplain and Hoe Terrace area		CC–M area
Flandrian		alluviation incision	W 3 *c.* 7.9 ka BP	
		Late Devensian interstadial Worthing Terrace	H *c.* 11.9 ka BP	
			BB 1 *c.* 16.5 ka BP	
	solifluction B	solifluction B WW–XX ponding		solifluction B
Middle/Late Devensian	incision	incision	icing *c.* 27.2 ka BP	incision
		RR in floodplain gravels	*c.* 31.7 ka BP	
	aggradation	aggradation (Hoe Terrace)	icing	M, CC lake formed in tributary *c.* 38.0 ka BP
	solifluction A aggradation	diversion east of Roosting Hill		
				incision
Early Devensian	interstadial			interstadial GG

pollen assemblage of section X and predate the first post-interstadial solifluction phase. Diapirs formed at the margin of depression T before the solifluction sheets were deposited. Involution occurred associated with the earliest solifluction sheet.

SECTION J–K–L

This section (figure 20) has been described in detail above. Following the interstadial (section K) a major diapir was formed in the upslope part of the section, before the covering solifluction diamicton was emplaced. This diamicton overlies disturbed red gravel interdigitating with grey silty clay at 20–25 m, and clayey gravel. The gravel is at a similar height to that in the section R–V, and could be part of the same aggradation, but here is affected by downslope solifluction associated with the final solifluction phase following incision.

SECTION N–P

This section (figure 31) has also been described in detail earlier. Disturbed gravel overlies interstadial sediments and is associated with grey clayey silt. The gravel and silt are at approximately the same level as the gravels of section R–V. The gravel may belong to the same post-interstadial aggradation, but has here collapsed into the subsidence hollow of VV, as well as being involved in diapirs. The grey clayey silt may be related to the silts seen in the early post-interstadial sequence of section R–V.

SUMMARY OF DEVENSIAN STRATIGRAPHY

It will be now useful to summarise the events in sequences described in this section. In three subsequent sections additional consideration of periglacial conditions, water levels and the regional terraces will be added to this outline.

Table 8 summarises the biostratigraphical sequence of the Devensian sections, and can be used as a framework for the succession. Table 10 summarises Devensian sedimentary events. The stadial herb pollen zone at section X underlies the interstadial soil at section X–V, and is the earliest Devensian biostratigraphic unit. Associated with the soil at section S–T–X are

74

interstadial limnic sediments in a depression, possibly caused by subsidence. The same interstadial is represented by the organic sediments in the depression at section D–E–F–G, where a much more complete sequence through the interstadial is present, following a herb pollen zone associated with the preceding stadial. Interstadial sediments are also present at section A, in a smaller depression with a less complete sequence, and at section N–P, where they are warped down as a result of post-interstadial subsidence. Pollen biozones with much *Calluna* pollen are associated with the interstadial soil at *c.* 23.5–25.5 m O.D. in sections V, GG and also K, where the sediment is much disturbed by a diapir.

Following the interstadial, in the M–CC area, downcutting took place, forming the depression in which stadial organic sediments were later formed. Further south, following the interstadial at section R–Z (figure 24) are fluviatile sediments, first stratified sand, later gravel, interbedded with solifluction diamictons. The herb pollen biozone at section R is within this sequence, which represents the west margin of an aggrading floodplain subject to solifluction, first from the east and south.

Formation of this floodplain was eventually followed by incision, with solifluction sheets forming downslope into the valley, but between the erosion base and the solifluction diamictons sequences of organic sediments and inorganic lake sediment were deposited. These are seen in the area north-west of Roosting Hill (figure 8, 9, 30, 34), with a thickness of *c.* 1.5 m (Bed F), and in the area to the south-west of Roosting Hill, where they are much thinner and are cut off by the solifluction diamicton on the slope (section WW–XX, figure 38). The origin of the pondings which led to the lake sediments is considered in a later section.

Incision in the Whitewater valley was followed by aggradation of gravels, containing channels with organic sediments with stadial herb floras. These gave radiocarbon ages of *c.* 31.7 ka BP, *c.* 27.2 ka BP and 11.9 ka BP.

CHAPTER SIX

The Flandrian (temperate) Stage or Post-glacial

The Flandrian began some 10 000 years ago and shows forest development as in the older temperate stages. This is seen best in the pollen record of many lake sites in East Anglia (e.g. Bennett, 1983, 1986), as well as in the sediments of the Broads of Norfolk and in Fenland. At Beetley, Flandrian lake sediments occur in the depressions on the Whitewater valley slopes and as peat and alluvium in the valley itself. Detailed pollen analytical results from a depression will be reported elsewhere.

Flandrian limnic sediments in the depression Q (figure 33) lie at a level above the floodplain at 24.4–25.8 m O.D., with deposition starting near the Late Devensian/Flandrian boundary. In the floodplain itself there are channels with Flandrian peat overlying the gravels and underlying alluvium. The peat reaches a thickness of 2.5 m and is covered by up to 1 m of alluvium. A pollen study of the peat and alluvium was made by Heptinstall (1984) at the northern end of the Whitewater valley (site W 3 in figure 2). The results showed that accumulation of peat began before the expansion of *Alnus* pollen frequencies at the beginning of Godwin's zone VIIa. A radiocarbon determination of *c.* 7.9 ka BP was obtained near the base of the peat. Heptinstall also studied a nearby small hollow with a spring, slightly raised above the floodplain level, which was interpreted as a ground-ice hollow (site W 1 in figure 2). Accumulation of organic sediment here started at the end of the Devensian late-glacial, as indicated by the pollen assemblages and a radiocarbon date of *c.* 9.7 ka BP.

The distribution of alluvium in the Whitewater valley is further considered in chapter 9.

CHAPTER SEVEN

Evidence of periglacial conditions

Periglacial phenomena provide important evidence for cold stage climates and conditions. Freeze–thaw alternations in the active layer and the development and degradation of permafrost give rise to identifiable sediments and structures. The processes concerned and the climates associated with them are found in arctic and subarctic regions today, so present-day processes can be linked with cold-stage processes. If, as at Beetley, periglacial sediments can relate to organic horizons dated by radiocarbon, a chronology of periglacial events can be suggested.

Sediments and structure relating to periglacial processes were found to be widespread in the Roosting Hill area. Most are of Devensian age; a few are from an earlier cold stage. In this section this evidence is considered together with the problem of the origin of irregular topography, which may be related to subsidence or thermokarst or both. The evidence for periglacial conditions lies in both localised features, such as thermal contraction cracks and involutions and those more widely spread, such as solifluction deposits and cover sands. The variable width shown by the Whitewater valley can also be related to periglacial processes.

Two background matters are important in relation to the following discussion. The first is the occurrence of springs and seepages on the east and west sides of the Whitewater valley (figure 2), and the second is the lower level of the Chalk/Pleistocene boundary in the immediate area, with a contour map of this boundary showing a closed depression below O.D. (Auton, 1982). Around this basin the Chalk surface rises rapidly to over 20 m O.D. and is covered by Pleistocene glacigenic sediments to a level of over 45 m O.D. on the surrounding plateaux. The springs and seepages thus reflect artesian pressures from the higher Chalk and Pleistocene reservoirs of the surrounds.

DEPRESSIONS

Shallow depressions, filled with water perennially or from time to time, are a feature of the Roosting Hill area. They may originate through melting of ground ice or icings, or through collapse after dissolution of the underlying Chalk, as discussed by West (1987). That collapse does occur in the area was shown by the overnight appearance in June 1984 of a 3-m deep hole 5 m across in a field at Spong Hill, 1 km north of Roosting Hill. In this area there is a considerable thickness of Pleistocene glacigenic sediments over Chalk; thus a borehole 0.5 km to the north-west of this collapse showed 43 m of Pleistocene sand and gravel over Chalk at −3.7 m O.D. (Auton, 1982).

Depressions are abundant in the eastern part of Whin Covert (e.g. site Q), and to the north of this area, where the depression at VV is the site of a spring and permanent standing water. Other depressions occur in the Whitewater valley, such as that at site W 1 described by Heptinstall (1984) near the confluence of the Blackwater and Whitewater. To the north of this confluence and the west of the railway there is a further area of irregular topography, marked H

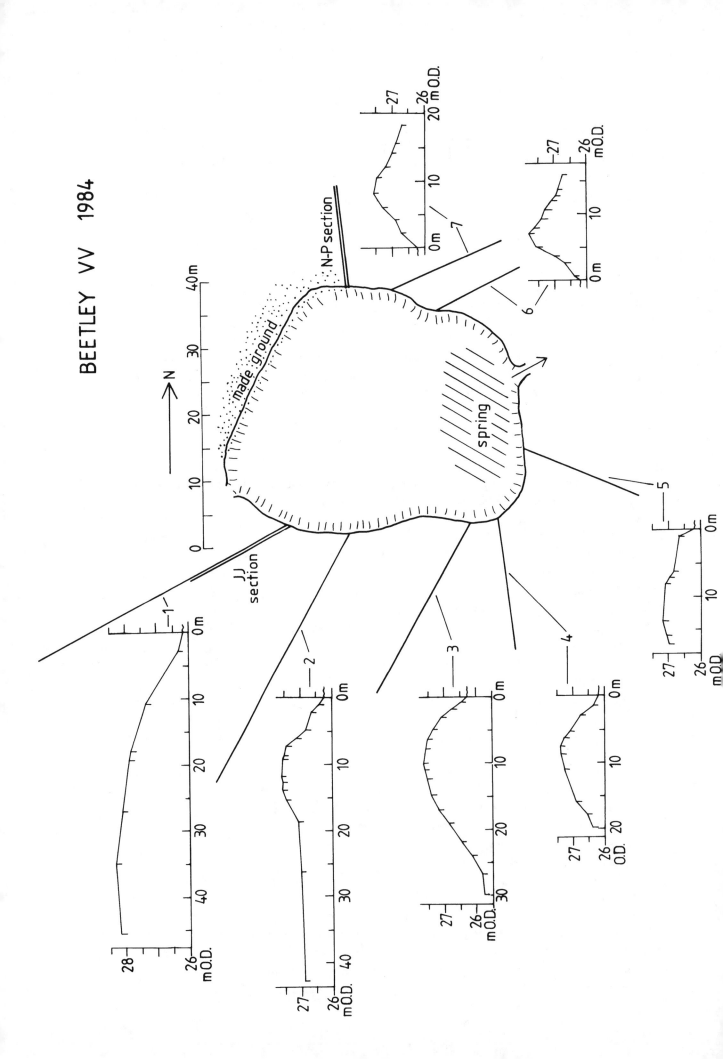

BEETLEY VV 1984

in figure 4; they are likely to have originated through the melting of ground ice. The hollows appear to be Late Devensian in age, since the lows are at heights down to 23.9 m O.D., near the level of the adjacent alluvium and low Devensian terraces. Further downstream, the alluvium in the area of the Swanton Morley gravel pits has much variation in level, perhaps related to collapse or ground ice.

In addition to these depressions seen at the surface, buried depressions of Devensian age occur, as already described for example in section R–V (figure 24).

As mentioned above, depressions may originate through collapse after dissolution or through the melting of ground ice. The origin of the partially rimmed depression VV following collapse has already been considered. This depression on the valley slope at VV appears to owe its origin to both collapse and ground ice. Figure 40 shows the shape of the depression and the contours of its immediate surrounds. To the east, on the downslope side, there is a clear rim, with steep interior walls. To the west the margin of the depression joins the surrounding slopes without a rim. There is a spring in the eastern part of the depression (also shown on the 1928 edition of the 1:2500 O.S. map).

Two trenches were excavated in the western margins of the depression, JJ to the south-west, N–P to the north-west (figures 31, 32). The N–P section gives clear evidence of collapse of the interstadial organic and overlying sediments, at a time certainly later than the formation of the interstadial sediments and the overlying sands and gravels. Collapse was sufficient to allow the deposition of Flandrian organic sediments in the depression. To the east, the presence of a marked rim round the depression (figure 40) suggests the former presence of a ground-ice mound (pingo) which sloughed off its covering sediment. Melting of the ice has left a form which is similar to decayed pingos of an open-system type, as described by Holmes *et al.* (1968) in central Alaska and Hughes (1969) in the Yukon, both in areas of discontinuous permafrost. Such open-system pingos are often the source of artesian springs, which may be perennial, the water derived via percolation in permafrost-free areas in elevated parts of the catchment. Spring water from VV has certainly dissected the gravel on the floodplain to the east.

The open-system pingos described by the above-named authors usually lie near the foot of slopes, adjacent to the valley floor, with a preference for south, south-east or east facing slopes, where permafrost may be less continuous. The VV depression is near the foot of a slope facing east (figure 2), so paralleling a common location of the open-system pingos described by Holmes *et al* and Hughes (*loc. cit.*). It was thus formed after the valley slope had started to form, perhaps coeval with the development of the surrounding solifluction slope, but after the period of gravel aggradation of post-interstadial times (see figure 38). The presence of a spring at present day implies an artesian source, probably associated with dissolution and collapse, as seen at the western end of the depression, and with growth of ground ice when permafrost developed in post-interstadial times. The morphology of the VV depression and the section in the N–P trench thus indicate a dual origin for the depression.

Two further depressions are seen on the east side of the valley in transect 14 (figure 43), associated with a present-day spring. These are also likely to be the result of ground-ice development. Irregular ground at the foot of slopes in the Whitewater valley is common (see transects in figure 43), and may likewise have arisen through the melting of ice associated with seepages.

INJECTIONS

This term is used to describe individual deformation structures, such as diapirs, not occurring in a regular patterned way as with involutions. They were seen in several Devensian sections, as follows, from north to south:

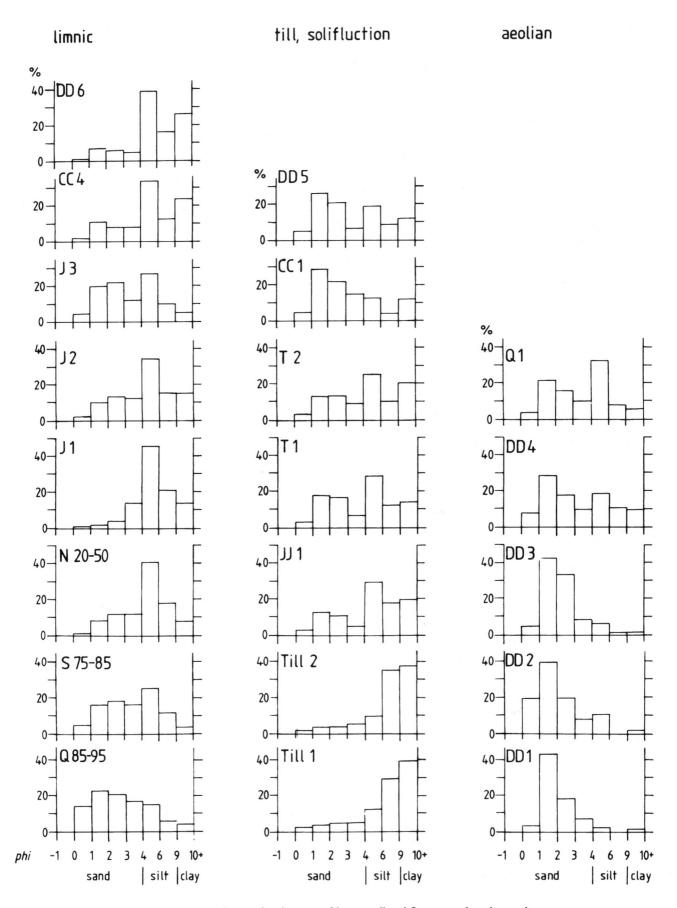

Figure 41. Particle-size distributions of limnic, till, solifluction and aeolian sediments.

Section HH to NN (figure 30)   Injections are later than the lacustrine sediments, rising into the overlying solifluction diamicton. At 17 m the underlying gravel rises abruptly, distorting the sequence of lake sediments. At 20 m these are faulted down next to an injection of gravel.

Section DD (figure 8) and EE (figure 9)   Injections of the marginal lacustrine sediments of Bed F into overlying solifluction sediments are seen along these sections.

Section J–K–L (figure 20)   An injection of sand at 26 m overturns older organic sediments at K, interstadial in age. Fawn chalky till is involved in this injection to the south at 30 m. The injection is earlier than the period of solifluction which formed the overlying sediments.

Section R–X (figure 24)   Deformations at 2.5 m, 18 m and 34 m are seen at the top of the underlying gravel. They appear to be earlier than the upper part of the overlying fluviatile sands, but involve the lower part of these sands. The balloon structure at 2.5 m resembles that at section L.

Section N–P (figure 31)   An injection of gravel at 9 m disturbs the organic sediments at section P. The injection post-dates the interstadial.

All the injections described above, except those at section DD, occur at the margins of basins or channels containing fills of relatively impermeable organic sediments or silts and clays. They appear to be structures related to the release of hydrostatic pressure. Such release, by its stratigraphical context, occurred in post-interstadial times but before a major period of solifluction in sections J–K–L and R–X. In sections HH–NN and DD injection occurred later, after the time when the lacustrine sediments (Bed F) associated with the Gramineae–Cyperaceae p.a.b. were deposited. Here it is associated with the beginning of the period of major downslope solifluction.

The origin of the structures may lie in the localisation of hydrostatic pressures resulting from the development or decay of permafrost. It is notable that both periods of injection precede or are associated with the development of solifluction sheets.

SOLIFLUCTION

Diamictons resulting from solifluction (gelifluction) in the Roosting Hill area appear to be all of Devensian age, except for the diamicton which lines the channel in gravels in which Ipswichian organic deposits occur in sections B and C (figure 16). This bed, of grey sandy silt and clay with angular and rounded flints, must be Wolstonian in age, resulting from a period of solifluction after the incision of the Wolstonian Beetley Terrace, but before the rise in water level which led to the aggradation of the Ipswichian organic sediments.

In contrast to the single section with undoubted Wolstonian solifluction, Devensian solifluction deposits are widespread. The upper western slopes of the Whitewater valley show a mantle of 1–2 m of stony sand and silt overlying the Hungry Hill Gravels, the junction between them affected by festooning. Presumably the mantle and festooning are a result of Devensian periglacial processes, though it might be expected that an older periglacial regime in the Wolstonian would have produced similar effects, concomitant with erosion and terrace formation in the valley.

In the northern part of the studied area, this solifluction mantle downslope overlies the lake sediments in the area of sections CC, DD, and M, and thins out towards the centre of the valley, as seen in section QQ (figure 13). Figure 41 shows particle-size analyses of the solifluction mantle upslope (sample DD 5) and over the lake sediments (sample CC 1). As expected, the sediments are ill sorted, and they differ from the finer diamictons of the southern part of the area described below.

To the south, the mantle becomes in places more silt- or clay-rich and gleyed on the lower slopes, where it overlies the channels and depressions containing organic sediments. Thus in section B and C (figure 16) grey stony silt overlies the interbedded pale sand and grey clay above the Ipswichian organic sediments.

Further south, as already described in section R–V, an early solifluction facies (A; table 10) alternates with a fluviatile facies within the depression, and a later solifluction sheet (B) descends over these sediments into the present valley (figure 38). Figure 41 gives particle-size analyses of sediments which are interpreted as solifluction diamictons in the southern area. The high proportion of fines is probably related to the upslope occurrence of clays, silts and tills associated with the Anglian gravels (figure 38). Particle-size distributions of dark grey till (till 1 in figure 41), fawn till (till 2) and two samples (T 1, T 2) of diamicton from the R–X section (figure 24) indicate the difference in composition between the silt-rich till and the post-interstadial diamictons.

In the section J–K–L (figure 20) the same later downslope solifluction sheet is seen. Here it overlies a red silty gravel, which may be interpreted as solifluction of the fluviatile gravels seen in section R–V. To the west and upslope of section J–K–L a single solifluction diamicton is seen in section D–G (figure 22) overlying the depression containing organic interstadial sediments, and becoming more silty to the east. In the section A (figure 22), again upslope, a single diamicton of sandy gravel with an irregular base overlies the interstadial organic sediments.

In section N–P (figure 31), sandy gravel passes laterally into grey silt, both overlying organic interstadial sediments. The gravel and silt are comparable to the fluviatile and solifluction facies of the section R–V. They are affected by collapse at the southern end of the section and by injection at the northern end. On the south-western side of the depression VV, the section JJ (figure 32) is very different, and shows a diamicton of brown stony silt with sand partings overlying a grey-blue diamicton with small chalk pebbles. The brown stony silt may be a solifluction diamicton, derived from upslope till. The lower grey-blue sediment diamicton may also originate by solifluction, since its particle-size distribution bears a similarity to the analyses of diamictons from section R–V but is unlike the till analyses, which are much richer in fines (figure 41).

There is localisation of the later solifluction cover in the southern area. In section Q (figure 33) it is absent in the form seen in section R–V and to the north, but cover sand is present. In the knoll at the eastern end of Whin Covert it is also absent, with coarse flint gravel directly underlying sandy soil. The localisation is probably related to the complexity of the upslope distribution of the underlying coarse and fine sediments.

The section R–V (figures 24, 28) shows that aggradation of gravels accompanies the beginning of solifluction at the western margin of the valley. Later, incision takes place and solifluction mantles the slope so formed. These changes must reflect changes in the stream load and discharge, perhaps associated with factors involving the development of permafrost and changes in load in the Middle/Late Devensian. Increased solifluction may result in increased load, but if permafrost develops strongly and the active layer is reduced in depth, load may decrease. Analysis of such changes, which are likely to accompany the onset and degradation of permafrost, will be of importance for clarifying cold stage climatic change.

The difference noted above in the solifluction sheets in the north and south of the area is a result of the constitution of the underlying upslope sediments, which are more silt-rich to the south. As a result solifluction appears to have started earlier in the south, so diverting the Ipswichian drainage to the east of Roosting Hill.

THERMAL CONTRACTION CRACKS AND COVER SANDS

Only four structures which may be interpreted as thermal contraction cracks have been observed in the Roosting Hill area. The oldest, shown in figure 16, underlies the stony silt and clay lining the channel which is filled by Ipswichian organic sediments, and which is interpreted as a solifluction diamicton. The wedge-like structure penetrates the underlying gravels for at least 2 m. It is probably Wolstonian in age, since it lies in a channel lower than the Wolstonian terrace gravels.

The remaining contraction cracks are associated with Devensian sediments. One underlies 1.2 m of solifluction sand and gravel on the slope north-west of section F–G (site Y; figure 2). It penetrates Hungry Hill gravels and is filled with sandy silt and clay of a composition similar to the silty facies of the diamicton described above, but without the stones. It is not possible to determine the age of this structure within the Devensian. It may be associated with the period of permafrost antedating the injection features discussed above, but certainly predates a period of solifluction. A second Devensian structure is seen in the section R–X (figure 24), at 1 m, adjacent to and limiting the bed of organic sediment at site X. This crack post-dates the organic sediment but predates the interstadial soil, and is therefore thought to be early Devensian in age. A third Devensian structure was described by Markham (1967) as penetrating the Ipswichian organic sediments in the area of the section B–C.

Two episodes of cover sand deposition are recorded in the sections. The older immediately underlies the solifluction mantle in the western part of the DD section (figure 8) in the northern part of the area studied, while the younger occurs at the surface over a much wider area.

The sand in the DD section is at the western margin of the depressions containing lake sediments (Bed F). It appears to have been deposited on the Wolstonian terrace gravels during the life of the lake or in the interval between that time and the beginning of the solifluction episode which resulted in the deposition of the overlying 2 m of diamicton. Particle-size analyses DD 1, 2 and 3 (figure 41) show well-sorted sand, with DD 2 coarser in grain size. DD 4 shows an admixture of fines and may represent accumulation in a wetter environment.

The younger sands overlie the solifluction mantle in the northern part of the area, not as a continuous visible sheet, but as pockets in involutions in the upper part of the mantle, e.g. at 40 m in section QQ (figure 13), 51 m in section DD (figure 8). To the south this sand becomes more widespread. Near the contraction crack at site Y a shallow basin with 20 cm of cover sand was seen on top of the solifluction mantle. Further south, near site T in the section R–X (figure 24) 20 cm of brown sand overlay the fine facies of the solifluction diamicton. At section N–P (figure 31) a bed of silty sand occurred at the surface, and at section JJ (figure 32) 60 cm of brown sand was seen over the diamicton along the length of the trench. A similar extensive brown sand with scattered flints was seen at the surface in the section Q trench (figure 33). A particle-size analysis of this sand (Q 1, figure 41) shows a bimodal histogram similar to that of a sand-loess, but the finer component of the cover sands in both sections N–P and Q may be derived from the underlying finer sediments by cryoturbation.

These younger sands extend over the gravel sheet to the east of the VV area. They are the final Devensian sediments deposited in this area and also on the western slopes of the Whitewater valley, with an age certainly later than the c. 27.2 ka BP radiocarbon determination at section XX (figure 38). Their best preservation is in the southern part of the area where the latest solifluction sheet is not so well developed. Possibly cover sand accumulation was partly contemporaneous with solifluction.

ICINGS

The gravels near the alluvial level south of transect 5 (figures 2, 7) are included in a terrace named the Hoe Terrace (see below). This terrace is dissected by channels now containing alluvium (figure 43). The margin of the Hoe Terrace is the rise associated with older terraces or solifluction slopes (e.g. figure 38). The Whitewater valley in this area is remarkable for variation in width. It widens rapidly north of the knoll at the east end of Whin Covert, is restricted opposite Roosting Hill, widens out again north of Roosting Hill and is restricted again at the confluence with the Blackwater, before widening out into the Wensum valley. These variations are seen well in figure 7 and the transects shown in figure 43.

The widenings of the Whitewater valley coincide with short tributaries now fed by springs, the positions of which are shown in figures 2 and 7. Thus on the west slope springs emerge from the decayed ground-ice mound VV and north-west of Roosting Hill, and on the east slope opposite Whin Covert (transect 14) and up a tributary valley to the east of here.

Such widenings of a valley floor are characteristic of valleys which have been subject to icing (See Hopkins et al., 1955; Protaseva, 1967; Kudriatseva 1981). The formation of river icings forces the spread of the river channels, resulting in wider braided systems. There is clear evidence that springs were flowing during cold climate time of the Devensian, as shown by the development of a ground-ice mound at VV and the stratified sediments associated with the valley north-west of Roosting Hill at sites CC and M, and further south in the transect 13 section.

A present-day situation similar to VV and its spring is described by Holmes et al. (1968), who report features which indicate icings in central Alaska, and which are associated with minor streams arising from pingos. The presence of springs in the Devensian would favour the development of seasonal, or perhaps perennial, icings which derive from perennial subterranean water sources, analogous to the icings associated with springs at the present time in the Arctic, for example those described by Liestøl (1977) in Spitsbergen and by van Everdingen (1981) in Northwest Territories, Canada.

The time of icing would include the time of formation of the Hoe Terrace. Since this is followed by dissection to produce the WW depression (figure 38) with a radiocarbon age of c. 27.2 ka BP of organic sediments at XX, a widening of the valley precedes this date.

It may be noted that ground-ice mounds are not present at all spring points at Beetley. Holmes et al. (1968) describe ground-water requirements for open-system pingo growth, including rates of flow, water and permafrost temperatures, and blockage of ground-water flow to localise sites for pingo formation. In permafrost areas spring discharges may be concentrated with resultant high discharge rates, so preventing freezing (French & Heginbottom, 1983). Evidently, such controls as these were satisfied in the case of the VV spring and the spring to the east on transect 14, but not in the M–CC area, where no depression was seen.

ORIGIN OF STADIAL LIMNIC SEDIMENTS AT SITES CC, DD, M, WW

The lacustrine sequence (Bed F), c. 1.7 m thick, of the DD, CC and M sections has already been described above. The Gramineae–Cyperaceae p.a.b. from the basal organic sediment of the depression, radiocarbon age c. 38.0 ka BP, indicates cold climate herbaceous vegetation in the area, presumably also the conditions under which the later silts and clays of the lake were also deposited. The lake sediments reach a height of about 27 m O.D., near the level of the older pre-Ipswichian terrace gravels to the north, suggesting that the height of these gravels controls the height of the lake. The aerial photograph in figure 36 gives an indication of the area covered by the lake.

The depression occupied by the lake is bordered on its west and north-west side by Ipswichian sediments (section DD; figure 8) and pre-Ipswichian terrace gravels (section EE; figure 9) and to the south by Anglian glacigenic gravels at section M (figure 34). The depression is either local, resulting from collapse in Devensian times, or is associated with a valley parallel to the Ipswichian valley but re-excavated later at a time of lower base level (see sections DD, GG–M; figures 8, 30) and draining north-east to the present valley. The extent, height and regular base of the sediments in the depression, and the regular way they overlap older sediments at the margin, make it unlikely that the depression is related to collapse. The second explanation requires a valley, possibly fed by springs, as at present, which supplied the stratified sands overlying the detritus mud, and which is subsequently dammed to give a depth of water of at least c. 1.7 m, so giving the fining-up sequence described previously (figure 35).

The damming may have occurred in the tributary or the main valley. Such damming could perhaps have taken place through blocking of the channel by solifluction. However, the stratigraphy indicates that strong solifluction followed the formation of the lake sediments, rather than preceding it. A more convincing explanation of lake formation is through damming of a minor short tributary by aggradation of sediments in the main valley. Damming of tributaries in this way has been described by Mackay (1963) in the Mackenzie delta area of Canada.

We may then envisage the following sequence of events. After downcutting of the main and CC–M valleys, gravels aggraded in the main valley only, since solifluction contemporaneous with the aggradation forced the drainage east of Roosting Hill (section R–Z). As a result the CC–M valley was ponded, leading to alluviation in this side valley starting at about 38 ka BP. Icings in the main valley may have assisted the ponding. Subsequently, the gravels in the main valley were dissected to give the Hoe Terrace seen today, with further ponding at about 27.2 ka BP (section WW, transect 13, figure 38), possibly associated with later icings.

It is remarkable that the possible terrace on the east slope (E. slope; figure 44) in the main valley north of transect 5 (figure 43) starts with a steeper gradient between 26 and 27 m O.D. just downstream of the point of confluence of the tributary with the main stream; it is not found upstream of this. The coincidence may possibly be explained by supposing that at a later stage an icing in the main valley blocked drainage, and that the stream discharge produced a surface grading strongly downstream. On the disappearance of the icing, the unconsolidated sediments of the CC–M tributary were quickly eroded to the base level of the main valley at the time, giving the shallow tributary valley with springs of the recent landscape.

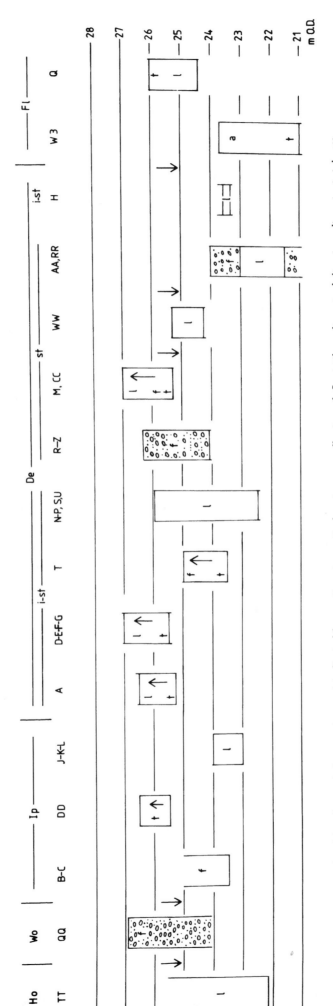

Figure 42. Water level heights and changes (arrows) indicated by sections in particular stages; a, alluvium; f, fluviatile sediments; l, limnic sediments; t, telmatic sediments.

CHAPTER EIGHT

The relation of the stratigraphy to water levels, hydroseres and palynology

The presence at Beetley of a series of organic and fluviatile sediments dating from a succession of stages gives the means of studying water level changes over a long period of time. Such an opportunity is available from no other site. There is also the opportunity to study related phenomena such as hydroseres and the origin or taphonomy of the pollen assemblages.

Thus a comparison of the many sections seen at Beetley leads to some general conclusions concerning changes of water level, hydroseres and pollen taphonomy. The sediment sequences in the basins of deposition, whether channels or depressions, reflect changes of water level at particular times, and these changes of water level are related to changes in the associated pollen spectra. The comparison of levels at different times is shown in figure 42.

The Hoxnian lake sediments indicate a lake level exceeding 25 m O.D., possibly by only a few metres. The Wolstonian aggradation recorded in the Beetley Terrace reaches 27.5 m O.D. The water level changes recorded in the sediments of the later stages are, however, more complex.

In the Ipswichian, Devensian and Flandrian there are two distinct types of channel or depression. Some are lined by solifluction deposits and others show an infill resting directly on underlying sands or gravels. The presence of unsorted solifluction deposits lining channels or depressions indicates that the related slopes were subject to solifluction before being filled with sediment at a time of higher water level. If they were water filled at the time of formation, or with slopes not subject to solifluction, the solifluction lining will be absent. There are also two types of infill: those of uniform sediment type, indicating infill over a time of high water level, and those in which there is a change of sediment resulting from a rise of water level, resulting in a reversed hydrosere. The following describes these variations.

IPSWICHIAN SECTIONS

B, C   The channel is lined by solifluction deposits. The earliest organic deposits of the channel filling include wood peat containing pine cones. Younger sediments are sandy muds and fluviatile sands. The loss-on-ignition results (table 3) reflect this change. A rising water table is indicated, flooding the valley peat and introducing pollen from a wider floodplain catchment, as seen in the pollen diagram of Phillips (1976). These changes indicate a rise in water table from c. 23.5 to 25 m O.D.

DD   The section indicates a further rise of water table in a later part of the temperate stage, flooding an organic soil on the Ipswichian valley slope at a height of c. 25.5–27.5 m O.D. Telmatic organic sediments were then formed to a height of c. 26.5 m O.D., rather than the more fluviatile sediments of section B–C. The pollen diagram shows this change clearly, as do the loss-on-ignition results from sections DDA and DD 3 (table 3).

J, K, L This section shows a shallow depression lying on sands and gravels. The sediment infill following depression formation is mainly inorganic with three more organic horizons. A fluctuating water table between *c.* 23 and 24 m O.D. is indicated, following formation of the depression in temperate stage times and the rise in water table indicated in section B–C.

DEVENSIAN INTERSTADIAL SECTIONS

A This section shows a small depression lined by a solifluction diamicton, indicating that the depression was formed before a relative rise in water table. The infill follows a later rise of the water table. The basal part of the filling is more organic, with high Cyperaceae pollen values. The later sediments are more inorganic and contain more inwashed pollen. The change indicates a rise in water table to above *c.* 26.5 m O.D., but is probably partly related to subsidence, with inwash replacing more telmatic sediments.

D–E–F–G This section shows a basin of organic sediment, with the lowest organic sediments not resting on solifluction deposits but showing a transition from the underlying pale sand, through brown mud with laminations to brown wood peat. The wood peat is succeeded by limnic sediment, which becomes more inorganic towards the top, where it includes a sediment derived by solifluction. The slumping of the organic sediments towards the centre of the basin, observed by J. Webb, indicates collapse at a time later than the basin fill. The sediment sequence seen in the basin, of a reversed hydrosere of wood peat followed by mud, indicates an early shallow pool in which telmatic wood peat then formed, with a later rise in water table. As with the section A, this rise may probably be partly related to subsidence, which certainly took place later, as indicated by the slumping. The absence of a solifluction lining indicates initial slow basin formation with immediate flooding, forming the transitional muds seen at the base of the filling. A relative rise of water table from *c.* 25.5. to 27 m O.D. is indicated by these changes.

T, GG These sections show reversed hydroseres in small shallow pools. At the base of the pool filling is *Carex* peat with high frequencies of Cyperaceae pollen. This is succeeded by more inorganic sediments associated with inwash and higher frequencies of *Calluna* pollen. In section T fluviatile sands follow, indicating a regional relative rise in water table and flooding of the land surface in the interstadial. In this section the reversed hydrosere is at *c.* 23.5–24 m O.D. with fluviatile sediments to *c.* 25 m O.D. At section GG the reversed hydrosere is at 25–25.8 m O.D.

N–P, S–U The sections show rather uniform silty organic limnic sediments resting on gravel, with no telmatic sediments at the base, which indicates a high water level from the start of sedimentation. If the pools originated through subsidence, as seems most likely, the depressions so formed were water filled from the start. The level of the limnic sediments is *c.* 22.5–26 m O.D.

DEVENSIAN STADIAL SECTIONS

The organic sediments considered here occupy situations different from the channels or depressions associated with Ipswichian or Devensian interstadial sediments described above.

AA, RR These sections show brown sandy muds in shallow channels in the floodplain gravels of the Whitewater. These are limnic sediments formed in abandoned channels of the braided floodplain. The water levels indicated, c. 21.5–23 m O.D., are those related to the floodplain at the time.

M, CC   These sections show thin telmatic organic sediments, later flooded by a rise in water table as a lake formed in their area. The rise indicated is from 25.4–27 m O.D. The origin of this lake is discussed above.

WW   This section shows shallow-water limnic organic sediments at a height of 24.3–25.3 m O.D. in a depression or channel cut in the underlying gravels.

R   The thin sandy mud of this section is a result of a pool associated with solifluction lobes when a high or perched water table was present, at c. 24.5 m O.D.

FLANDRIAN SECTION

Q   The section through this present depression, with sedimentation starting in the Devensian late-glacial at 24.4. m O.D., shows limnic sediments resting directly on gravel. The depression evidently contained water from the start of sediment accumulation, with the age of the infilling indicating subsidence to form the depression before the end of the Devensian.

CARBONATE-RICH SEDIMENTS

The Hoxnian sediments are rich in calcium carbonate, probably derived largely from spring discharges following the retreat of the Anglian ice. No such sediments were found in the Ipswichian or Devensian, and the drainage in these stages may have been more superficially derived from weathered Anglian and Wolstonian gravels. In the Flandrian marls are known in the district related to springs, and tufa mounds occur, as at Badley Moor, East Dereham.

SUMMARY OF WATER LEVEL CHANGES

Rises in water level are recorded in the early and later parts of the Ipswichian (sections B–C, DD) and in the interstadial (sections T, GG), though the latter may be relative and related to subsidence. If the depressions described above are caused by subsidence, then subsidence, resulting in the formation of telmatic and limnic sediments, occurred in the Ipswichian (section J–K–L), before the Devensian interstadial (section A), near the beginning of the Devensian interstadial (section F–G), during the Devensian interstadial (sections T, GG, N, P, S–U), and in the Late Devensian (section Q). Water-level heights described above, and periods of lower level resulting in erosion, are summarised in figure 42.

The Hoxnian, Ipswichian, Devensian and Flandrian levels overlap, indicating a remarkable stability in temperate and interstadial times, with downcutting of a Wolstonian aggradation before the Ipswichian and in post-Devensian interstadial times before the Middle and Late Devensian floodplain gravels were deposited. These changes reflect a stability of the landscape and hydrology of the area around the Whitewater valley during the more temperate periods of the Middle and Late Pleistocene.

(figure 43)

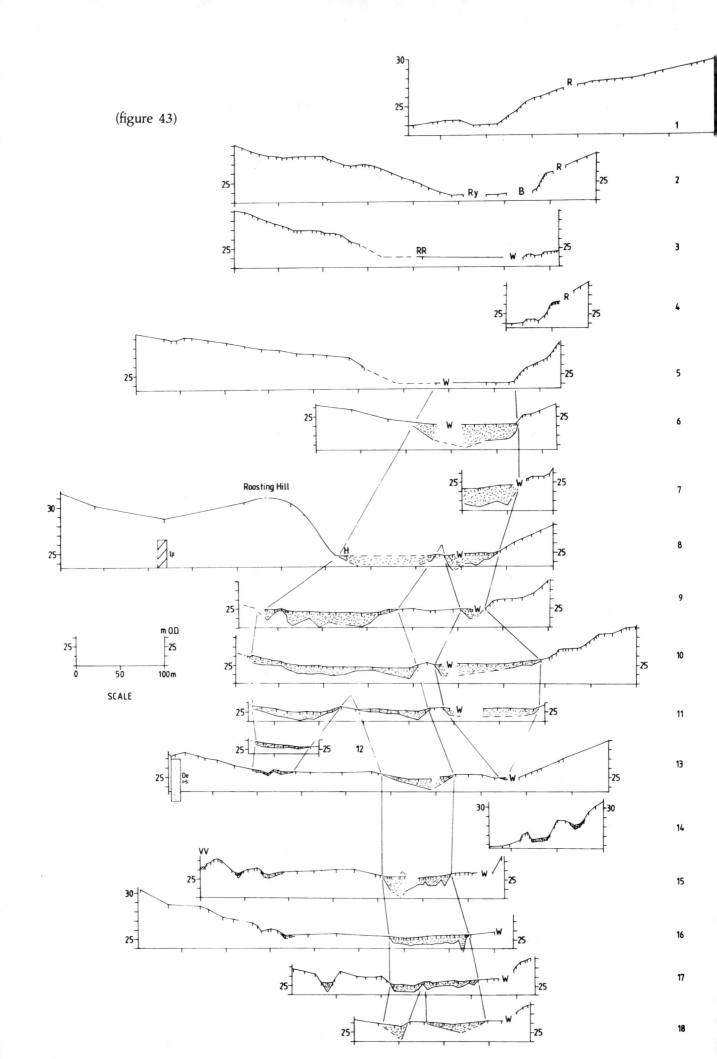

CHAPTER NINE

The history of the Whitewater, Blackwater and Wensum valleys

The abundance of evidence at Beetley for the relative and radiocarbon dating of sediments and their contained floras gives the means of reconstructing in some detail the history of the valleys of central Norfolk over the period of time since the ice advance of the Anglian (cold) Stage. Aggradations of fluviatile sediments can be allocated to particular periods in cold stages, so enabling a sequence of aggradations in the valleys to be developed. The identified aggradations are not associated with temperate stages, during which lake or quiet-water sediments were formed.

Since the gravels forming the Beetley Terrace seen in the QQ section (figure 13) can be allocated to a cold stage (Wolstonian) between the Hoxnian and Ipswichian (temperate) Stages, the opportunity arose to trace this relatively dated terrace into the Wensum valley, and to relate it to other terraces observed locally. Several transects (figure 43) were first levelled across the Whitewater and Blackwater valleys. Their positions are shown in figures 1 and 2. Terrace flats underlain by sand and gravel were then levelled downstream past the confluence of the Wensum and Blackwater as far as Castle Farm, Swanton Morley. The positions of the levelled terraces are shown in figure 1. Long profiles of the Whitewater, Blackwater and Wensum alluvium and terrace levels were then drawn (figure 44).

The Beetley Terrace is traceable into the Wensum valley at least to Castle Farm. There are indications of terraces higher than the Beetley Terrace in the area; these may belong to Anglian retreat times or to older parts of the Wolstonian. They were not examined further. Lower terraces are also present in the Whitewater and downstream. Certain of these are demonstrably Devensian in age.

Distinguishing Wolstonian and Devensian terrace gravels upstream of section R–Z (see figure 2) becomes impossible without biostratigraphic control, since the levels are nearly similar, as seen in figure 42 and 44. Thus the gravels in section Q (figure 33) and those forming the knoll at the east end of Whin Covert reaching 26.6 m O.D. (transect 17) could, by their level, be either age. Upstream of transect 6 (figure 2) the lowest terrace gravels are dissected and it becomes impossible to trace the lowest terrace in the southern part of the studied area. The transects 8–19 (figure 43) show this dissection.

In addition to the Beetley Terrace, two further and lower terraces of wide distribution have been distinguished; these are named the Hoe and Worthing Terraces. A further problematic feature resembling a terrace and with a steep gradient is seen at 1200–1500 m downstream from transect 5 (E. slope; figure 44).

Figure 43. West–east transects in a sequence upstream across the Blackwater (1,2) and Whitewater valleys (3–18), with a south–north transect (19) in the southern part of the study area. Positions of transects shown in figures 1 and 2. B, Blackwater stream; W, Whitewater stream; R, road; Ry, railway. Levels of Ipswichian and Devensian interstadial sediments are shown on transects 8 and 13 respectively. Flandrian peat and alluvium hatched, with areas of Flandrian alluviation connected between transects.

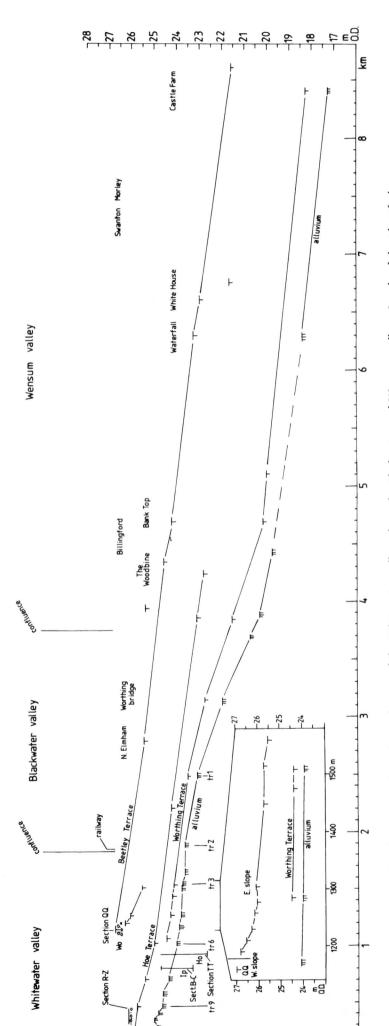

Figure 44. Long profiles downstream from the study area of the Whitewater valley, along the Blackwater and Wensum valleys. Levels of the identified terraces and of the alluvium are shown.

The Beetley terrace type section is the profile and section QQ (figure 13). The height of the sand and gravel reaches 26.8 m O.D., with the flat above at about 27 m O.D., 3 m above the nearby alluvium level. Transect 5 (figure 43) shows the relation between the two. Upstream, a flat of the same age may be present in the transects 10 and 16 (figure 43). Downstream a flat near the same level is seen in the transect 3 across the eastern part of Beetley Common, and possibly in transect 2 to the north of the Blackwater. Further downstream an extensive terrace at 25.6 m occurs at the eastern end of North Elmham village, dividing the present Blackwater and Wensum valleys. The profile of the terrace can be extended further downstream past terraces south of Billingford, Waterfall Cottage and White House, Swanton Morley, to an extensive terrace east of Castle Farm, on the south side of the Wensum and 4.5 m above the nearby alluvium level. A pit in this terrace showed 3 m of massive sandy angular flint gravels with sand lenses covered by 30 cm of sand with scattered flints (cover sand). It is likely that the area to the north of the river and opposite the large meander at Swanton Morley is also a part of the same terrace, here subjected to solifluction cover as in the western part of the QQ section at Roosting Hill.

The Beetley Terrace is clearly deeply dissected, at least in part before the Ipswichian, as shown by the section at Roosting Hill. The presence of a Wolstonian terrace in the Wensum parallels the occurrence of terraces ascribed to the same cold stage in the Waveney valley (Coxon, 1979; Coxon et al., 1982) and the Nar valley (Ventris, 1986).

A lower terrace, named the Hoe Terrace, occurs in the Whitewater valley upstream of transect 5, reaching 26.2 m O.D. in transect 15. Possible correlates of this level were found downstream to Billingford. In the area between transects 8 and 18, this level of gravels is dissected by the main Whitewater and by the tributaries emerging from springs to the west (site VV) and east, as shown in figure 43. On the slope to the west of this area, in the sections R–Z (figures 24, 28), gravels at a similar height (24.0–26.4 m O.D.) have been shown to be Devensian post-interstadial. They likewise predate a period of dissection, since they are succeeded by solifluction on the slope to the east, as seen in figure 38. If these gravels belong to the same aggradation, as here interpreted, then the Hoe Terrace is post-interstadial in the Devensian, but earlier than the date of 27 ka years given by the radiocarbon date at site XX on the western slope of the valley (figure 38). It thus appears to be late Middle Devensian in age. If the aggradation of this terrace blocked the tributary of the CC–M area to the north, as already suggested, a lower limit of age of c. 38.0 ka BP is given to the aggradation.

The lowest terrace of the Whitewater, named the Worthing Terrace, is seen downstream from transect 5, where it is slightly above the alluvium level. In this area sections in it were seen at sites AA, BB and RR (figure 13). The terrace can be traced downstream, past St Margaret's Church, Worthing, and Home Farm, Worthing, through Billingford Common, the area where gravels at a low level overlie Ipswichian organic sediments (Coxon et al., 1980), and to Castle Farm, where it is a clear feature 1 m above the alluvium level. The age of the terrace is probably later than the downcutting of the Hoe Terrace, i.e. later than c. 27.2 ka BP. The age of the sands and gravels underlying this terrace in the Whitewater covers a long span in the Middle and Late Devensian, indicated by the radiocarbon determinations from the sites RR (c. 31.7 ka BP), BB (c. 16.5 ka BP) and H (c. 11.9 ka BP), proving the complexity of aggradation and erosion in the valley.

In addition to these three terraces a feature resembling a terrace occupies a short stretch on the eastern side of the Whitewater downstream of transect 5, commencing opposite the QQ section and ending near the W 1 hollow at the north (E. slope, figure 44). The height of the southern part of the feature is near 27 m O.D. sloping rather quickly to 25.7 m or so to the north. There is no clear relation to other terrace features upstream or downstream, though isolated flats occur intermediate in height between the Beetley Terrace and the lowest terrace downstream. The origin of this possible terrace is considered above in relation to a possible icing in the valley.

The long profile of the alluvium level of the Whitewater, Blackwater and Wensum valleys is also shown in figure 44. Changes in the slope are seen at the narrowing of the valley near the confluence of the Whitewater and Blackwater and at the confluence of the Blackwater and Wensum. In the area between transects 6 and 18 the depth of alluvium was mapped by probing to the underlying gravel, with the results shown in figure 43. Spring flow from the west (site VV) and the east, together with the main Whitewater flow, has dissected the low (Hoe) terrace. The dissection must be later than *c.* 11.9 ka BP, the radiocarbon determination from the Devensian late-glacial at section H, but earlier than *c.* 7.9 ka BP, the radiocarbon determination from near the base of the alluvial sequence at site W 3. The latter date and a pollen diagram from the same site (Heptinstall, 1984) indicate that peat formation and alluviation started in the lower part of the Whitewater in Godwin's Zone VI, blanketing a complex dissected surface of gravel.

The terrace distribution described above indicates that the present valleys were superimposed on the Hoxnian landscape in Wolstonian times.

CHAPTER TEN

Beetley in the context of the East Anglian Pleistocene

In this chapter the salient features of the Pleistocene history of the Whitewater valley are related to the wider Pleistocene of East Anglia. At Beetley the relation between the Hoxnian lake sediments and the underlying outwash gravels shows clearly that the latter, the Hungry Hill Gravels of Phillips (1976), are Anglian in age, and thus associated with the chalky boulder clay of the area. The landscape left by the retreat of the Anglian ice was irregular, with lake basins. The detailed study of the lake basin at Hoxne, Suffolk, the type site of the Hoxnian temperate stage, shows a similar irregularity (West, 1956). In both areas the surface of the glacigenic deposits under the lake sediments resembles dead-ice topography. Other occurrences of Hoxnian lake sediments occur in steep-sided valleys, as at Mark's Tey, Essex (Turner, 1970), possibly tunnel valleys. Hoxnian lake sediments may occur in well-developed present-day valleys, as at Beetley and Barford (Phillips, 1976) or on higher plateaux, as at Hoxne and Athelington, Suffolk (Coxon, 1985). It is reasonable to assume that, as in the Flandrian, little topographical change took place during the Hoxnian temperate stage in inland areas.

There is no evidence for glaciation in the Beetley area during the succeeding cold stage, the Wolstonian. During this stage, however, it is clear that the Hoxnian drainage system was transformed to the system we see today. In some areas the drainage ways developed in closer relation to the Hoxnian drainage, and less so in other areas, giving the variable relation of Hoxnian lakes to today's landscape. Thus in the Wolstonian the major valleys carried aggradations of gravel now seen as terraces, as in the Whitewater and Wensum and also in the Nar valley in western Norfolk (Ventris 1986) and in the Waveney valley in central East Anglia (Coxon, 1979).

Ipswichian sediments in East Anglia are found in the valleys established during Wolstonian times, usually overlying fluviatile gravels, as at Beetley and Swanton Morley, or on older solid rocks, such as the Gault at Wretton in the Wissey valley of west Norfolk (Sparks & West, 1970). They are covered by Devensian fluviatile gravels, except for the Beetley section B–C, where Ipswichian sediments occupy a channel which was later abandoned. A wide variety of environments is indicated by the sediments of this temperate stage in East Anglia and their contained floras, including a sandy fluviatile facies, backswamp fens, reedswamp, and open-water lacustrine conditions. The high *Pinus* pollen frequencies in the early part of the temperate stage (Ip I) contrast with the low frequencies seen in the early Hoxnian (Ho I), and are likely to be associated with the poor soils which were developed in the area on Anglian outwash gravels weathered through the Hoxnian and Wolstonian.

A remarkable feature of the vegetation history of the Ipswichian Stage, well developed at Beetley, is the occurrence of high non-tree pollen spectra associated with remains of large grazing vertebrates. This association has been discussed by Phillips (1974), Stuart (1976, 1986) and Turner (1975). The pollen spectra are characterised by high frequencies of Gramineae pollen and unusually high frequencies of pollen of Compositae Liguliflorae and *Plantago lanceolata*, a combination thought to indicate meadow grassland when found in the late

Flandrian. In the Ipswichian it seems likely to indicate meadow associated with the expansion of floodplain communities and grazing, and has been recorded in Ip II and Ip III.

Evidence for Devensian environmental history in East Anglia is found within fluviatile sediments which form low terraces in the valleys and in the near-surface sediments of higher ground, where periglacial structures such as stripes and polygons are well preserved. Features associated with ground-ice occur in valleys and on higher ground, often in areas where springs occur today. Beetley has an exceptionally rich record of Devensian environments, showing both stadial grassland and interstadial forest vegetational history, now associated with depressions on the valley slopes, with fluviatile floodplain gravels and with solifluction sediments. The interstadial and stadial floras at Beetley are found in a variety of facies, including open-water lacustrine sediments, sedge fens, shallow fluviatile sediments, and a fluviatile drift mud facies. As already discussed, this has allowed closer study of variations in fossil content allied to taphonomy,.

The Beetley interstadial sediments occur in depressions on the west slope of the valley and show a vegetational history more complete than is seen in the similar interstadial assemblages recorded at Chelford, Cheshire (Simpson & West, 1958) and Wretton (West et al., 1974). The interstadial at Beetley is therefore correlated with the Early Devensian Chelford interstadial, as first suggested by Phillips (1976). Stadial grassland pollen spectra preceded the interstadial and also follow it, as at Wretton (table 8). They occur in organic sediment (radiocarbon age c. 38.0 ka BP) below the major solifluction sheet which seals earlier Devensian sediments on the present valley slope. This solifluction sheet thus appears to date from a later part of the Devensian, suggesting that the surface periglacial features associated with solifluction elsewhere in East Anglia are also of later Devensian age. Within the floodplain gravels organic sediments with similar stadial pollen spectra occur at c. 31.7 ka BP and c. 16.5 ka BP, with a further organic sediment dated to the late-glacial at c. 11.9 ka BP.

The study thus gives considerable detail to the landscape history of an upper minor tributary of the R. Wensum, relating it to the nature of the underlying Anglian and older sediments and to the vegetational history of the Middle and Late Pleistocene.

References

Auton, C.A. (1982). A preliminary study of the sand and gravel deposits of part of central Norfolk (1:25 000 sheets TF 91, 92 and TG 01). Institute of Geological Sciences, Keyworth.

Behre, K.E. (1981). The interpretation of anthropogenic indicators in pollen diagrams. *Pollen et Spores*, **23**, 225–45.

Bennett, K.D. (1983). Devensian late-glacial and Flandrian vegetational history at Hockham Mere, Norfolk, England. I. Pollen percentages and concentrations. *New Phytologist*, **95**, 457–87.

Bennett, K.D. (1986). Competitive interactions among forest tree populations in Norfolk, England, during the last 10 000 years. *New Phytologist*, **103**, 603–20.

Blake, J.H. (1888). *The geology of the country around East Dereham*. Memoirs of the Geological Survey, England and Wales.

Coope, G.R. & Pennington, W. (1977). The Windermere Interstadial of the Late Devensian. *Philosophical Transactions of the Royal Society of London, B*, **280**, 337–9.

Coxon, P. (1979). *Pleistocene environmental history in central East Anglia*, Ph.D. Dissertation, University of Cambridge.

Coxon, P. (1985). A Hoxnian interglacial site at Athelington, Suffolk. *New Phytologist*, **99**, 611–21.

Coxon, P., Clarke, M.R., Horton, A. & Wilcox, D.W. (1982). The Waveney Valley. In *Quaternary Research Association Field Guide: Suffolk, May 7–9, 1982*, ed. P. Allen, Part 3, 1–35.

Coxon, P., Hall, A.R., Lister, A. & Stuart, A.J. (1980). New evidence on the vertebrate fauna, stratigraphy and palaeobotany of the interglacial deposits at Swanton Morley, Norfolk. *Geological Magazine*, **117**, 525–46.

Ehlers, J., Gibbard, P.L. & Whiteman, C.A. (1987). Recent investigations of the Marly Drift of northwest Norfolk, England. In *Tills and Glaciotectonics*, ed. J.J.M. van de Meer, pp. 39–54. Balkema: Rotterdam-Boston.,

French, H.M. & Heginbottom, J.A. (eds) (1983). Guidebook to permafrost and related features of the northern Yukon Territory and Mackenzie Delta, Canada. Fourth international conference on permafrost, Fairbanks, Alaska.

Gibbard, P.L. & Aalto, M.M. (1977). A Hoxnian interglacial site at Fishers Green, Stevenage, Hertfordshire. *New Phytologist*, **78**, 505–23.

Godwin, H. (1975). *History of the British Flora*. 2nd ed. Cambridge University Press: Cambridge.

Groenman-van Wateringe, W. (1986). Grazing possibilities in the Neolithic of the Netherlands based on palynological data. In *Anthropogenic Indicators in Pollen Diagrams*, ed. K.-E. Behre, pp. 187–202. Balkema: Rotterdam.

Hall, A.R. (1980). Late Pleistocene deposits at Wing, Rutland. *Philosophical Transactions of the Royal Society of London, B*, **289**, 135–64.

Heptinstall, S. (1984). *Late Devensian and Flandrian vegetational history of a tributary of the River Wensum, Worthing, Norfolk*. M.Phil. Dissertation, University of Cambridge.

Holmes, G.W., Hopkins, D.M. & Foster, H.L. (1968). Pingos in Central Alaska. *U.S. Geological Survey Bulletin*, **1241-H**.

Hopkins, D.M., Karlstrom, N.V. *et al.* (1955). Permafrost and ground water in Alaska. *United States Geological Survey Professional Paper* **264-F**.

Hughes, O.L. (1969). Distribution of open-system pingos in central Yukon Territory with respect to glacial limits. *Geological Survey of Canada, Paper*, **69–34**.

Kudriatseva, V.A. (1981). *Merzlotovedenie.* Moscow University Publishing House. (in Russian).

Lambrick, G. & Robinson, M. (1988). The development of floodplain grassland in the Upper Thames Valley. In *Archaeology and the Flora of the British Isles*, ed. M. Jones, pp. 55–75. Oxford University Committee for Archaeology Monograph No. 14.

Liestøl, O. (1977). Pingos, springs and permafrost in Spitsbergen. *Norsk Polarinstitutt Arbok 1975*, 7–29.

Mackay, R. (1963). The Mackenzie delta area, N.W.T. *Canada Geographical Branch Memoir*, **8**.

Markham, R. (1967). Interglacial beds at Beetley, Norfolk. *Ipswich Geological Group Bulletin*, **3**, 4–5.

Phillips, L. (1974). Vegetational history of the Ipswichian/Eemian interglacial in Britain and continental Europe. *New Phytologist*, **73**, 589–604.

Phillips, L. (1976). Pleistocene vegetational history and geology in Norfolk. *Philosophical Transactions of the Royal Society of London*, B, **275**, 215–86.

Protaseva, I.V. (1967). *Aeromethods in Geocryology.* Moscow: Akademia Nauk. (in Russian).

Reis, O.M. (1931). Die Oberbayerischen Seen. In *Handbuch der vergleichenden Stratigraphie Deutschlands: Alluvium*, ed. J. Stoller, pp. 178–86. Gebrüder Borntraeger: Berlin.

Sagar, G.R. & Harper, J.L. (1964). Biological Flora of the British Isles. No.95. *Plantago major L., P. media L.* and *P. lanceolata L. Journal of Ecology*, **52**, 189–221.

Shackleton, N.J. & Opdyke, N.D. (1973). Oxygen isotope and palaeomagnetic stratigraphy of equatorial Pacific core V28–238: oxygen isotope temperatures and ice volumes on a 10^5 year and 10^6 year scale. *Quaternary Research*, **3**, 39–55.

Simpson, I.M. & West, R.G. (1958). On the stratigraphy and palaeobotany of a late-Pleistocene organic deposit at Chelford, Cheshire. *New Phytologist*, **57**, 239–50.

Sparks, B.W. & West, R.G. (1959). The palaeoecology of the interglacial deposits at Histon Road, Cambridge. *Eiszeitalter und Gegenwart*, **19**, 123–43.

Sparks, B.W. & West, R.G. (1970). Late Pleistocene deposits at Wretton, Norfolk, I. Ipswichian interglacial deposits. *Philosophical Transactions of the Royal Society of London*, B, **258**, 1–30.

Stuart, A.J. (1976). The history of the mammal fauna during the Ipswichian/Last Interglacial in England. *Philosophical Transactions of the Royal Society of London*, B, **276**, 221–50.

Stuart, A.J. (1986). Pleistocene occurrence of *Hippopotamus* in Britain. *Quartärpalaeontologie*, **6**, 209–18.

Turner, C. (1970). The Middle Pleistocene deposits at Mark's Tey, Essex. *Philosophical Transactions of the Royal Society of London*, B, **257**, 373–440.

Turner, C. (1975). Der Einfluss grosser Mammalier auf die interglaziale Vegetation. *Quartärpalaeontologie*, **1**, 13–19.

Tutin, T.G. *et al.* (1964–80). *Flora Europaea.* Cambridge University Press: Cambridge.

van Everdingen, R.O. (1981). Morphology, Hydrology and Hydrochemistry of Karst in Permafrost Terrain near Great Bear Lake, Northwest Territories. *National Hydrology Research Institute, Paper*, **11**.

Ventris, P.A. (1986). The Nar Valley. In *The Nar Valley and North Norfolk, Field Guide*, ed. R.G. West & C.A. Whiteman, pp. 7–55. Quaternary Research Association: Coventry.

West, R.G. (1956). The Quaternary deposits at Hoxne, Suffolk. *Philosophical Transactions of the Royal Society of London*, B, **239**, 265–356.

West, R.G. (1980). Pleistocene forest history in East Anglia. *New Phytologist*, **85**, 571–622.

West, R.G. (1987). Origin of small hollows in Norfolk. In *Periglacial processes and landforms in Britain and Ireland*, ed. J. Boardman, pp. 191–4. Cambridge University Press: Cambridge.

REFERENCES

West, R.G. (1989). The use of type localities and type sections in the Quaternary, with especial reference to East Anglia. In *Quaternary Type Sections: Imagination or Reality?* ed. J. Rose & C. Schlüchter, pp. 3–10. Balkema: Rotterdam.

West, R.G., Dickson, C.A., Catt, J.A., Weir, A.H., & Sparks, B.W. (1974). Late Pleistocene deposits at Wretton, Norfolk, II. Devensian deposits. *Philosophical Transactions of the Royal Society of London, B,* **267**, 337–420.

Appendix I

NOTE ON TECHNIQUES

Pollen preparations were carried out according to the schedules of the Subdepartment of Quaternary Research, Cambridge (West, 1977). Macroscopic plant remains were extracted from samples, usually of 200 g, broken down over several days in a solution of dilute NaOH, then washed gently through a set of sieves, apertures 500 μm and 150 μm. The two resulting fractions were each scanned and macroscopic plant remains extracted.

Sections recorded in the later part of the field investigations were levelled to O.D. Those recorded in 1964 and 1972 were not, and their heights, shown bracketed in the section drawings, were estimated by interpolation from lines levelled downslope across the excavations at a later date.

REFERENCE

West, R.G. (1977). *Pleistocene Geology and Biology*. 2nd edn. Longman: London.

Appendix II

NOTES ON PALAEOBOTANY

Plant nomenclature follows *Flora Europaea* (Tutin *et al.*, 1964–80). In the identification of the cold stage macroscopic plant remains, the works of Bell (1968, 1970) have proved invaluable, especially their discussion of the problems of identification of particular taxa. The taxonomy of many cold stage genera is complex and incomplete, a fact perhaps related to the history and migrations of the cold stage flora, and much more needs to be done to develop our knowledge of cold stage floras, both from the fossil and the modern taxonomic points of view.

***Arabis* type**   In the sample BB 1 (Late Devensian) three flattened and rounded Cruciferae seeds were present. They are most similar to *Arabis* and *Cardamine*, the seeds of which are difficult to separate, especially in the fossil material. The largest and most complete fossil seed measured 1.7 × 1.3 mm and at its apex showed a curved radicular protrusion. Some of the marginal cells of the three seeds tend to be square, arranged in rows parallel to the seed margins, and most have raised margins. These features are more readily seen in *Arabis* spp. Not all *Arabis* seeds are necessarily winged (Tutin *et al.* 1964–80). The fossil seeds appear to be unwinged and although they do not match any native *Arabis* species, their general morphology most closely resembles *Arabis*. Bell (1970) describes two Devensian *Arabis* type seeds, though her drawing of these does not match those from Beetley.

Calluna vulgaris   The *Calluna* pollen percentages are higher where the sediments are more inorganic, as described in the discussions on sections A, T, V, Z and GG. The increase is ascribed to inwash of sediment as a source of the pollen, rather than to aerial origin. The detailed work by Peck (1974) on pollen dispersal in the Oakdale catchment in North Yorkshire, a catchment which contains a wide spread of *Calluna* heath, assists in this interpretation. In this catchment, which includes a stream and lakes, in waterborne pollen spectra *Calluna* pollen percentages were 10–16% of total determinable pollen, but only 1–3% of the local air catch. Peck also compared pollen representation with percentage representation of Callunetum in the upper Oakdale catchment. Callunetum covered 77% of the upper catchment, but *Calluna* only accounted for 21% of the annual stream pollen load, 17% of the annual lake pollen load and 6% of the annual air pollen load. It appears that in such a catchment higher frequencies of *Calluna* are associated with stream flow, a situation comparable to the proposed overland derivation of high *Calluna* pollen frequencies at Beetley.

    Calluna pollen is abundant in sediments of the interstadial, indicating the presence of stable heath communities and associated soil pollen bank. The pollen became widely distributed by slopewash and fluviatile dispersal. Later in the Devensian such redistribution is not seen. Heath is not recorded in the pollen diagrams, and earlier soils had been destroyed, probably by solifluction.

Cf. *Glyceria fluitans* Abundant Gramineae caryopses were recorded in sample X 30–60 (Early Devensian). Following the key of Körber-Grohne (1964), these caryopses were found to be similar to those of *G. fluitans* in their size, length and position of hilum, and structure of pericarp layers. However, the key does not cover all British genera of grasses. Not all fossil caryopses showed all these features, and on the basis of modern reference material it proved difficult to separate *G. fluitans* from *G. declinata*. The identification is tentative.

Juncus As described previously, certain macroscopic assemblages show high frequencies of *Juncus* seeds, but little else. These samples generally contained a proportion of degraded seeds which could not be identified in addition to those which were determined specifically. Such high frequencies of *Juncus* seeds are recorded in assemblages from Midlands Iron Age and later organic 'meadow' sediments, but with many other species (Lambrick & Robinson, 1988). It seems likely that changing water levels associated with the Beetley sediments concerned led to the destruction of much discrete organic material, except for the more resistant epidermis of the *Juncus* seed. In this connection, it is relevant to note that Milton (1939, 1948) has described the abundance of *Juncus* seeds surviving in various soils, particularly those of wet upland areas (see also Lazenby, 1955).

Primula L. subgenus *Aleuritia* (Duby) Wendelbo Sample BB 1 (Late Devensian) yielded six seeds ascribed to this taxon. The nomenclature follows *Flora Europaea* (Tutin *et al.* 1964–80). The section includes the native species *P. farinosa* and *P. scotica*, both plants with a restricted and more northerly distribution in Britain. Bell (1968) comments on the problem of comparing past and present *Primula* sect. *Farinosae* distributions and the associated difficulty of nomenclature.

Rumex L. subgenus *Acetosella*; *R. acetosella* group This taxon, recorded in sample BB 1, AA 40–60 and AA 70–90 (Devensian), includes *R. angiocarpus*, *R. tenuifolius* and *R. acetosella*. Within this group nuts of *R. angiocarpus* may be distinguished by the adhering perianth segments, as seen in the modern reference material. In fossil samples this characteristic may be less secure. Only two of the BB 1 nuts appear to have fragmentary pieces of perianth attached, while the nuts from the other two samples are poorly preserved. Consequently the *R. acetosella* group taxon is considered appropriate.

Salix In the Anglian late-glacial pollen spectra *Salix* pollen reaches the highest values seen at Beetley. It is associated with *Hippophae*, and probably represents shrub species. On the other hand, in the Devensian *Salix* pollen is rarely represented. The highest frequencies are in the late-glacial (Section H), where it is associated with tree birch pollen. In the Late Devensian sections showing high non-tree pollen percentages and abundant *Salix herbacea* leaves (sections BB), no *Salix* pollen is found. This tallies with observations indicating that pollen of dwarf *Salix* species is very much underrepresented in the pollen rain (Birks, 1973; Pennington, 1980).

Saxifraga L. sect. *Sedoides* Gaudin. *S. hypnoides/rosacea* In sample BB 1 (Late Devensian), abundant saxifrage leaf remains most closely resemble *S. hypnoides* and *S. rosacea*, and to a lesser extent *S. cespitosa*. *Flora Europaea* (Tutin *et al.* 1964–80) notes the following points relevant to identification to a specific level. First, in *S. rosacea* subsp. *rosacea* the leaves have '. . . obtuse, acute or slightly mucronate lobes'; such characters were seen in modern reference material and some fossil leaves. However, other fossil leaves have a tendency to show apiculate-aristate lobes, an attribute of *S. hypnoides*. Secondly, following *Flora Europaea*, *S. rosacea* usually has five-lobed leaves, while the most complete fossil leaves are either entire or three-lobed. The difficulty of separating *S. hypnoides* and subsp. *rosacea* in the fossil material is increased by possible confusion with *S. rosacea* subsp. *sponhemica* which has '. . . leaves with strongly mucronate or apiculate narrow lobes'. Webb (1950) describes the taxonomic difficulties of the dactyloid saxifrages of north-western Europe. The identification in sample BB 1 follows the nomenclature of Clapham, Tutin & Warburg (1962), suggesting that the fossil material belongs to section *Sedoides*, probably *S. hypnoides* and/or *S. rosacea*. Webb (1950) comments that species in the section are more or less drought-susceptible chamaephytes that hybridise readily among themselves.

REFERENCES

Bell, F.G. (1968). *Weichselian glacial floras in Britain*. Ph.D. Dissertation, University of Cambridge.

Bell, F.G. (1970). Late Pleistocene floras from Earith, Huntingdonshire. *Philosophical Transactions of the Royal Society of London, B*, **258**, 347–78.

Birks, H.J.B. (1973). Modern pollen rain studies in some arctic and alpine areas. In *Quaternary Plant Ecology*, ed. H.J.B. Birks & R.G. West, pp. 143–68. Blackwell Scientific Publications: Oxford.

Clapham, A.R., Tutin, T.G. & Warburg, E.F. (1962). *Flora of the British Isles*. 2nd edn. Cambridge University Press: Cambridge.

Körber-Grohne, U. (1964). *Probleme der Küstenforschung im südlichen Nordseegebiet. 7. Bestimmungschlüssel für subfossile Juncus-Samen und Gramineen-früchte*. August Lax: Hildersheim.

Lambrick, G. & Robinson, M. (1988). The development of floodplain grassland in the Upper Thames Valley. In *Archaeology and the Flora of the British Isles*, ed. M. Jones, pp. 55–75. Oxford University Committee for Archaeology Monograph No. 14.

Lazenby, A. (1955). Germination and establishment of *Juncus effusus* L. *Journal of Ecology*, **43**, 103–19.

Milton, W.E.J. (1939). The occurrence of buried viable seeds in soils at different elevations and on a saltmarsh. *Journal of Ecology*, **27**, 149–59.

Milton, W.E.J. (1948). Buried viable seed content of upland soils in Montgomeryshire. *Empire Journal of Experimental Agriculture*, **16**, 163–77.

Peck, R.M. (1974). *Pollen transport and deposition in Oakdale, north Yorkshire*. Ph.D. Dissertation, University of Cambridge.

Pennington, W. (1980). Modern pollen samples from west Greenland and the interpretation of pollen data from the British Late-glacial (Devensian). *New Phytologist*, **84**, 171–201.

Tutin, T.G. *et al.* (1964–80). *Flora Europaea*. Cambridge University Press: Cambridge.

Webb, D.A. (1950). A revision of the dactyloid saxifrages of north-western Europe. *Proceedings of the Royal Irish Academy*, **53B**, 207–40.

Appendix III

NON-MARINE MOLLUSCA FROM BEETLEY, NORFOLK

R.C. Preece

Department of Zoology, Downing Street, Cambridge

Three samples were analysed for non-marine Mollusca (table A1), two dating from the pre-temperate substage (Ho I) of the Hoxnian and one from the Late Devensian.

The grey silt (sample T) and white marl (sample TT) from sediments adjacent to section TTC both of Ho I age, were very shelly but the shells were badly comminuted and represented mostly by juveniles or broken fragments. Both samples produced similar assemblages dominated by *Bithynia*, but selective sorting is indicated by two facts. First, the number of opercula far exceeds the number of *Bithynia* shells, and second, the scarcity of bivalves (e.g. the complete absence of *Pisidium*) is striking. Bivalves and gastropods with wide apertures (e.g. *Lymnaea*) are generally scarce in strandline flood debris compared with gastropods with small apertures. The relative scarcity of *Lymnaea* in these samples may therefore also be significant. Sample T yielded a typical late-glacial facies with both *Gyraulus laevis* and *Armiger crista*, an association characteristic of lacustrine faunas at the end of cold stages that may, as here, persist into the early part of the ensuing temperate stage (cf. Kerney, 1977).

The assemblage from sample BB 2 comes from a silt lens laterally equivalent to a horizon of drift mud at section BB 2, radiocarbon-dated at *c.* 16.5 ka BP. It comprises a limited fauna of 13 aquatic species characteristic of lowland calcareous streams. The diversity is far lower than one would expect in such environments during temperate stages, but is richer than most faunas

Table A1. *Frequency of non-marine Mollusca from Beetley*

	T	TT	BB 2
dry weight (g)	350	400	500
Valvata cristata Müller	2	3	1
Valvata piscinalis (Müller)	15	13	9
Bithynia tentaculata (L.) shells	61	77	—
Bithynia opercula	194	466	—
Lymnaea peregra (Müller)	2	—	2
Gyraulus laevis (Alder)	4	—	1
Armiger crista (L.)	5	—	3
Sphaerium corneum (L.)[a]	2	3	6
Pisidium amnicum (Müller)[a]	—	—	6
Pisidium casertanum (Poli)[a]	—	—	10
Pisidium milium Held[a]	—	—	10
Pisidium subtruncatum Malm[a]	—	—	39
Pisidium hibernicum Westerlund[a]	—	—	4
Pisidium nitidum Jenyns[a]	—	—	57
Pisidium pulchellum Jenyns[a]	—	—	2

Note:

[a] number of valves

from an equivalent type of environment from the coldest parts of the Late Devensian. Bivalves of the genus *Pisidium* dominate. The occurrence of *P. pulchellum*, a relatively scarce species virtually unknown from British cold stage deposits, is particularly noteworthy.

The importance of this assemblage lies in the fact that it falls within a period for which there are virtually no other records. Kerney (1977) even questioned whether any Mollusca survived in England through the period of maximum cold of the Late Devensian between 26 ka and 14 ka BP. However, Holyoak (1982) noted that the organic deposits at Barnwell Station, radiocarbon-dated to 19.5 ka ± 650 a BP (Coope, 1968, 1980) had previously yielded a molluscan fauna with seven aquatic and five terrestrial species (Kennard & Woodward, 1922). Three of these aquatics are also in the Beetley assemblage. There are no other British faunas reliably dated to *c* 16.5 ka BP. A broadly similar aquatic fauna has been described from a shell marl filling a kettle-hole at Kildale, Yorkshire (Keen, Jones & Robinson, 1984) but there are suspicions that the basal radiocarbon date of 16 713 ± 340 a BP, based on moss fragments separated from the marl, has been seriously affected by 'hard-water error'.

REFERENCES

Coope, G.R. (1968). Coleoptera from the 'Arctic Bed' at Barnwell Station, Cambridge. *Geological Magazine*, **105**, 482–6.

Coope, G.R. (1980). The climate of England during the Devensian Glacial Maximum: evidence from fossil Coleoptera. *Quaternary Newsletter*, **30**, 11–13.

Holyoak, D.T. (1982). Non-marine Mollusca of the Last Glacial Period (Devensian) in Britain. *Malacologia*, **22**, 727–30.

Keen, D.H., Jones, R.L. & Robinson, J.E. (1984). A Late Devensian and Early Flandrian fauna and flora from Kildale, north-east Yorkshire. *Proceedings of the Yorkshire Geological Society*, **44**, 385–97.

Kennard, A.S. & Woodward, B.B. (1922). The Post-Pliocene non-marine Mollusca of the east of England. *Proceedings of the Geologists' Association*, **33**, 104–42.

Kerney, M.P. (1977). British Quaternary non-marine Mollusca: a brief review. In *British Quaternary Studies: Recent Advances*, ed. F.W. Shotton, pp. 31–42. Clarendon Press: Oxford.

Appendix IV

RADIOCARBON DETERMINATIONS FROM BEETLEY, NORFOLK

V.R. Switsur

Subdepartment of Quaternary Research, University of Cambridge

From the complex Devensian stratigraphy of the Whitewater valley, Beetley, five horizons yielded sediments containing residual organic material suitable for radiocarbon age determinations. The locations of the sections are shown in figure 2 and indicated by the symbols BB 1, H, M, RR, with one section, XX, being adjacent to the WW site on transect 13. In addition, radiocarbon determinations were made on two samples of organic sediment from Flandrian deposits at the northern end of the valley at W 1 and W 3.

THE SAMPLES

The provenance and nature of the samples were as follows:

Q-2292. Figure 13, section H. Coarse detritus mud overlying coarse flint gravel in wide shallow channel.

Q-2293. Figure 13, section BB 1. *Salix* leaves separated from laminated drift mud and silt overlying coarse sand and gravel, filling a shallow pond associated with the floodplain.

Q-2451. Figure 13, section RR. Brown sandy mud in a 4-m wide channel in floodplain.

Q-2294. Figure 34, section M, north-west of Roosting Hill. Coarse detritus mud deposited in shallow water in low part of irregular surface of black flint gravels.

Q-2717 and Q-2718. Figure 38, section XX, south-west of Roosting Hill. Silty mud deposited in depression in incised surface of floodplain gravels.

Q-2419. Figure 2, site W 3. Humified peat from channel in the floodplain underlying a metre of alluvium.

Q-2418. Figure 2, site W 1. Humified detritus mud from a possible ground ice hollow with a spring.

SAMPLE PREPARATION AND ANALYSIS

Organic deposits of any antiquity are likely to be contaminated with materials of different ages. Even small quantities of younger carbon compounds with their higher radioactivity can adversely affect the apparent age of a sample and it is important that, as far as possible, these are removed. The deposits investigated here, however, were well sealed and it is unlikely that recent carbon would have been incorporated, so that extensive pretreatment was not necessary. The extracted muds were all washed carefully with dilute hydrochloric acid and distilled water. Further pretreatment was only performed in the case of samples Q-2717 and Q-2718. Here the alkali-soluble and alkali-insoluble fractions were determined individually.

Similarly, in the cases of the humified samples, Q-2418 and Q-2419, the humic-acids fraction was extracted with a phosphate-buffered alkali solution and the determination was made of the ages of the purified humic acids.

The organic material was oxidised either in the high pressure combustion bomb (Switsur, 1973; Switsur & West, 1973) or the quartz tube to carbon dioxide, which, after purification, was converted through the pathway of lithium carbide and acetylene to benzene. This was used to prepare an accurate scintillation cocktail using a butyl-PDB scintillator. The activity of the carbon in the benzene was measured carefully in a fine-tuned liquid scintillation spectrometer dedicated to age determinations. In the case of the older, smaller, samples the background activity of the individual counting vials, of low potassium content glass, was measured both before and after the sample measurements, since the sample activity was so low. Measurements of the international oxalic acid standard activity and that of small known age samples were also made for comparison.

Conventional radiocarbon ages for the samples were calculated and the results are given in the table below. They are based on the Libby half-life for the radiocarbon isotope of 5568 years and the zero year of AD 1950. Corrections were made for the usual laboratory variables and the sample isotopic fractionation. The ages may be converted to the new, more likely, half-life of 5730 years by multiplication by the factor 1.03.

The uncertainties of the ages were calculated by the proper combination of the one standard deviation statistics associated with the counting of the sample, the background, and the international oxalic acid standard, and hence represent a 68% probability that the age lies within the stated range. It should be noted that for ages up to approximately three half-lives the uncertainties are reasonably symmetrical and their value may be doubled to obtain the 95% probability uncertainties. Beyond this point, however, because of the logarithmic nature of the decay curve, the uncertainties become less symmetrical and such a simple procedure is not possible. It is then necessary to return to the statistics associated with the activity measurements in order to calculate the 95% or other probabilities of the age uncertainty. This is most important as the limits of the age range of the method are approached, in order to decide whether or not there is any statistical difference between two given ages or whether any particular determination represents an infinite age.

RESULTS

See Table A2.

The samples Q-2717 and Q-2718 represent different fractions of the same sample; it will be seen that at the 68% probability quoted the ages are not in agreement, but at 95% probability

Table A2. *Radiocarbon ages*

Laboratory reference	Site reference	Radiocarbon age BP (ka)	Uncertainty +	Uncertainty −
Q-2419	Worthing W 3–1	7.88	0.09	0.09
Q-2418	Worthing W 2–4Q	9.86	0.11	0.11
Q-2292	Beetley H	11.92	0.08	0.08
Q-2293	Beetley BB 1	16.50	1.70	1.40
Q-2717	Beetley XX	26.19	0.58	0.54
Q-2718	Beetley XX	28.12	0.46	0.44
Q-2451	Beetley RR	31.75	0.89	0.80
Q-2294	Beetley M	38.00	1.20	1.00

(Q-2717: $+1.20/-1.05$ ka; Q-2718: $+0.95/-0.85$ ka) there is no statistically significant difference.

Table 10 above gives a concise summary of the sequence of Devensian and early Flandrian events and their timing.

REFERENCES

Switsur, V.R. (1973). Combustion Bombs for Radiocarbon Dating. *Proceedings of the 8th International Conference on Radiocarbon Dating, Lower Hutt, Wellington, New Zealand*, ed. T.A. Rafter & T. Grant-Taylor, **1**, 13–24.

Switsur, V.R. & West, R.G. (1973). University of Cambridge Natural Radiocarbon Measurements XII. *Radiocarbon*, **15**, 534–44.

Index

The only plant taxa indexed are those discussed in some detail